PLANT-ENVIRONMENT INTERACTION

Prof. Pravin Chandra Trivedi (b. 3 March, 1953), Ph.D., Post-Doct. (U.S.A.), F.L.S. (London), F.B.S., F.P.S.I., F.B.R.S., F.M.A., F.E.S., F.N.R.S., F.I.A.T., Professor, Department of Botany, has a brilliant academic career standing first in all examinations. Prof. Trivedi served as Head, Department of Botany and Co-ordinator, P.G. Course in Biotechnology (2003-06); Director, College Development Council (2001-04); Vice-Principal, Maharaja's College (2000-03) and is Member of Senate, Academic Council and Convenor of BOS Botany in the University of Rajasthan.

He worked as Post-Doctoral Fellow at the Department of Plant Pathology, North Carolina State University, Raleigh, U.S.A. in 1983-84. Prof. Trivedi has 36 years of teaching and research experience in the University of Rajasthan, Jaipur. He has published more than 250 research papers and review articles in journals of repute. He has edited and written more than 130 books and guided 40 Ph.D. students. He has been the Principal Investigator of 21 major research projects funded by national organisations. He is an elected Fellow of nine reputed academic bodies. Dr. Trivedi visited several major research laboratories and delivered invited talks in many international conferences.

Prof. Trivedi is an elected Secretary of the Indian Botanical Society (2005-2010); President, International Society for Conservation of Natural Resources; Vice-President, Mendelian Association; Council Member of Indian Science Congress Association, Kolkata (2007-09) and a member of the editorial and advisory boards of eight reputed journals. For his research contribution he was awarded Scientist of the Year Award (2001); M. J. Narsimhan Award (2004), Bioved Fellowship Award (2005), Birbal Sahni Foundation Honour (2007), Rashtriya Gaurav Award (2007), Distinction Award for Contribution in Life Sciences (2009) and Shiksha Vibhushan Samaan (2009).

The major scientific contribution of Dr. P.C. Trivedi falls in the field of Plant Pathology and Nematology. He and his students contributed a lot on nematode diseases of various vegetables and crops in the state of Rajasthan and their management using biocontrol practices and reported many resistant varieties after screening of germplasm. Dr. Trivedi has transferred technology from the lab to the field for the management of nematode pests. There practical solutions have been utilized by industry for large scale production of biopesticides. He has organised 20 National seminars and participated in more than 200 National and International conferences.

Prof. Trivedi served as chairman/member of selection & inspection committees for various senior positions in Education in India & also serving as panel of experts for evaluation and monitoring of research proposals for funding by D.S.T., U.G.C. and C.S.I.R., N. Delhi. He served as President, Indian Phytopathological Society, Central Zone (1997 and 2003); Convener, Jaipur chapter, Indian Science Congress Association, Kolkata (1999-2007); Councillor, Indian Botanical Society (1998-2001); President, Society for Promotion of Plant Sciences (2005-2006); Elected Council Member, Indian Science Congress Association, Kolkata (2003, 2005, 2007-2009); Member, Governing Body of Vardhman Institute; Member, Academic Council, Mahaveer Open University, Kota (2006-2009); Jaipur, Member, Rajasthan Board of Secondary Education, Ajmer (2010-2013).

Dr. Trivedi is actively working to promote scientific temper and popularize science among masses. Through Science Education Centre, Jaipur (U.G.C.), he developed a substantial good number of models, experiments & teaching aids to generate interest towards science in school students & masses. He actively participated in programmes on scientific issue on Doordashan and All India Radio and regularly writing on various issues pertaining to science and society and related to plants in the local news dailies and magazines.

PLANT-ENVIRONMENT INTERACTION

Editor

Prof. P. C. Trivedi

Ph.D., Post Doct. (U.S.A.), F.L.S. (London), F.B.S.,
F.P.S.I., F.N.S.I., F.E.S., F.N.R.S., F.M.A., F.I.A.T.
Department of Botany
University of Rajasthan
Jaipur-302055, India

Pointer Publishers

Jaipur 302 003 (Raj) India

Distributed by
Prem C. Bakliwal for
Aavishkar Publishers, Distributors
807, Vyas Building, Chaura Rasta
Jaipur 302 003 (Raj.) India
Phone : 0141-2578159
e-mail : aavishkarbooks@hotmail.com

ISBN 978-81-7132-642-6

2021

Published by
Mrs. Shashi Jain for
Pointer Publishers
807, Vyas Building, Chaura Rasta
Jaipur 302 003 (Raj.) India
Phone : 0141-2578159
e-mail : pointerpub@hotmail.com
Visit us at : www.pointerpublishers.com

Price : ₹ 2500.00

Printed at
Sheetal Printers
Jaipur (Raj.)

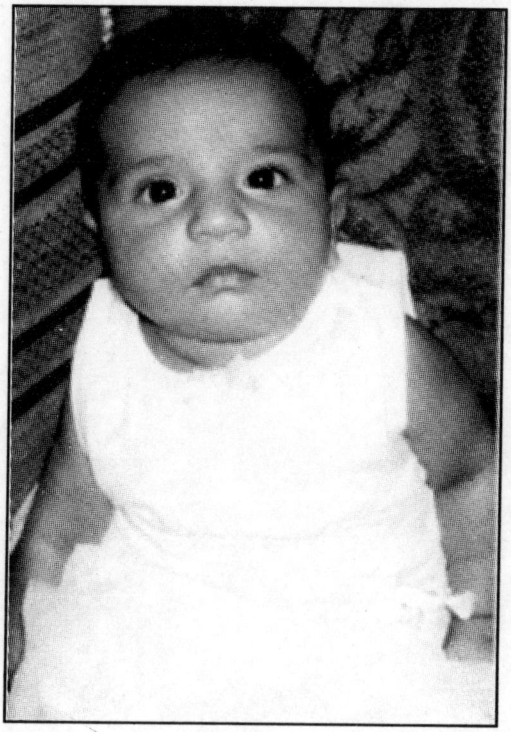

With Love and Affection to
Little Angel, Kyna

PREFACE

Before the dawn of human civilization, some 10,000 years ago, the earth boasted of a rich forest mantle and open woodlands of some 6,200 mha which, due to various development activities, for example, Agriculture, Industries, River Valley Projects, Roads, Buildings and Communication have shrunk to some 4,200 mha by the turn of the century. Forest plays a crucial and critical role in the maintenance of global ecological balance and in the operation of carbon, nitrogen, oxygen and water cycles of the biosphere, vital for agriculture and for the sustenance of life on earth.

Plants absorb the deadly cosmic rays; absorb the greenhouse gas carbon dioxide and some other harmful gases; intercept millions of tones of dust; serve as acoustic screens against noise pollution and neutralize the massive heat build up on earth due to solar radiation. As the foster mother of agriculture; they bring rain; feed the rivers and streams with water for irrigation; hold the soil in site; check soil erosion and maintain the soil mantle of the earth; influence the local and regional climate, form the watersheds in the mountainous region and provide the necessary inputs of agriculture like timber for tools and implements; fibre for rope, fodder for liverstock & fertiliser for crops and above all conserve the valuable "gene pool" of the wild relatives of the crop plants upon which much depends the ruture of agriculture.

As modern industrial society has developed over the past century, the environmental health risk arising from human developmental activities—hazardous wastes and pollution, global warming & stratospheric ozone depletion—has increased to several fold. The Federal Center for Disease Control and Prevention in the U.S. places air pollution related health costs at US $ 14 billion a year (UNEP Report 2001). Allergies, cancers and threats to reproductive health are on the rise everywhere but most in the U.S. and the European union. The pace of environmental destruction is not in years, not in months and not in days, but alarmingly every hour and minute. The UNEP Report (1996) said—'And while the Rio Earth Summit (June 5-15, 1992) was on :

- 600 to 900 plant and animal species become 'extinct'.

- 1,97,256 hectares of arable land turned into 'desert'.
- 5,34,000 hectares of 'tropical rainforest' were 'distroyed'.
- 3.3 million people (consumers) were added.

All over the world, destruction of forests (plants) in beginning to turn the ordinary seasonal weather into economic and ecological disasters. The recurring cyclones which kills thousands almost every year in the Asian region especially in the Philippines, Bangladesh and South India; the flood which perishes several cattle heads and uproot people every year in north India, Pakistan and Bangladesh are all results of global deforestation. The tsunami disaster of December 2004 has vindicated the protective role of mangrove forest and coral reefs in the coastal region. Reports have confirmed that countries, which had kept their mangrove forest intact and had healthy coral reefs were less badly affected than similar ones where they had been destroyed (UNEP Report, 2005).

The Present book "Plant-Environment Interaction" incorporates critical review articles on all major aspects of this subject in 16 chapters. The authors have tried to synthesise up-to-date information, which is pertinent for researchers and students alike. Chapters included are : High temperature stress response in crops; Response of cyanobacteria to UV radiation; Assessment of quaternary vegetation & climate of Kumaun Himalaya and Kaleidoscopic behaviour of *Spirulina* under different culture conditions. Chapters on Plant extracts—A tool for genotoxic damage repair; An approach to identify different varieties of *Mangifera indica* with the help of morphological traits; Biology & economics of *Prosopis cineraria*; Phloem of some common ferns & effect of human interferences on orchid diversity provide detailed information on the subject. Articles on Phytoastrology : A Botanical approach to astrology; Concurrent growing of green manure crops in rice; Indian Forests and Research—Past & Present; Comparative suitable nutrient solution for the growth of plants and its effect on calcium deficiency and Microbial endophytes : The untapped resources for bioactive extrolites have added to the value of the book.

This volume will definitely serve not only as an excellent reference material but also as a guide for teachers, research workers, students & workers in the field of plant sciences.

I bow in thankfulness before Lord Almighty for blessing my daughter Priyanka with a little angel, Kyna on 20th April, 2010 Tuesday (4.16 am) at Atlanta (U.S.A.) in the presence of Nani Kusum Trivedi and Papa-Kumar Gaurav. By core of my heart I appreciate beautiful little doll, Kyna for her cherishing smile which fills life in me. At this happy moment, we the family members—P.C. Trivedi and Kusum (Nana-Nani) and Rohit-Neha (Mama-Mami) pray the All Merciful God to bestow upon Kyna dream filled future and good health.

I am thankful to all the learned contributors for their co-operation in compiling useful information for this book. Heartfelt thanks go to these contributors, each of whom has endeavoured to present up-to-date section of his area of expertise & has given willingly of very valuable time and knowledge. The opinion and text contained herein are those of the authors and I have tried to honour their ideas in the original shape. I am sure the information given in this book will prove helpful to those interested in plant-environment interaction studies.

I record my sincere gratitude to my wife Kusum, Priyanka–Kumar Gaurav, Rohit-Neha for their consistent support in my academic and scientific endeavors and encouragement during the preparation of this book. I appreciate the all round co-operation and support of Mr. Vipin Jain of Pointer Publishers, Jaipur for printing and publishing this book with patience, care and interest.

Jaipur **Prof. P.C. Trivedi**

CONTENTS

Chapter 1

High Temperature Stress Response in Crops

D.S. Gill and Tilak Raj

INTRODUCTION

Stress is often defined as the rise in temperature beyond a threshold levels for a period of time sufficient to cause irreversible damage to plant growth and development. In general, a transient elevation in temperature, usually 10-15°C above ambient, is considered heat stress. However, temperature stress is a complex function of intensity (temperature in degrees), duration and rate of increase in temperature. The extent to which it occurs in specific climate zone depends on the probability and period of high temperature occurring during day and night. Temperature stress is generally defined as the ability of the crop plant to grow and produce economic yield under high temperatures. However, while some researchers reported that night temperature are major limiting factors other have believed that day and night temperatures do not effect the plant growth and development independently and that the diurnal mean temperature is a better predictor of plant response to high temperature with a day temperature having a secondary role (Peet and Willits, 1998). Temperature due to high ambient temperature is a serious threat to crop productions world wide (Hall, 2001). Rising temperature may lead to altered geographical distribution of crops by allowing the threshold temperature for the start of the season and crop maturity to research earlier (Porter, 2005). At moderately high temperature injuries or death may occurs only after long term exposure. Direct injuries include proteins denaturation and aggregation, and increased fluidity of membrane lipids. Indirect temperature injuries include inactivation of enzymes in chloroplast and mitochondria, inhibitions of proteins synthesis, protein degradation and loss of membrane integrity (Howarth, 2005). The tolerance conferred by HSPs results in

Department of Botany, Punjab Agricultural University, Ludhiana – 141004, India.

improved physiological phenomenon i.e. photosynthesis, assimilate partitioning, water and nutrient use efficiency, and membrane stability (stability (Camejo *et.al.*, 2005; Ahn and Zimmerman, 2006; Momcilovic and Ristic, 2007).

TABLE 1

Comparison of temperature limits of growth in different plants groups

S. No.	Group	Features	Upper temp. limit (°C)
1	Psychrophiles	Plants thrive at cold temperataure (0-20°C), are freezing stress tolerant. Plants of tropical and sub-tropical regions are susceptible to cold temperature.	15-20
2	Poikilotherms	Plants attain the low temperature of their environment, while still alive. Chilling sensitive plants are injured at temperature above freezing point upto 15°-20°C and chilling resistant plants can grow at temperature near 0°C. Temperate species are frost tolerant.	15-20
3	Mesophiles	Plants show maximum growth at moderate temperature (10°-30°C). Plants in the natural habitat as well as plants cultivated in tropical and sub-tropical regions.	35-45
4	Thermophiles	Plants thrive at high temperature (heat stress):	
		(a) Poor	
		* Vascular plants	45
		* Mosses	50
		(b) Moderate	
		* Algae	55-60
		* Fungi	60-62
		(c) Extreme	
		* Blue green algae	70-73
		* Photosynthetic bacteria	70-75
		(d) Chemotrophic bacteria	>100
		(e) Heterotrophic bacteria	>100
5	Poikilohydrous/ Resurrection plants	Plants have protoplasmic resistance to injury under extreme dehydration; assume metabolic activity on rehydration within relatively short time. Spores, seed and sporophytes of moss and grasses.	0-100

Source : Bewley (1979), Brock (1985), Gaff (1986), Ingram and Bartel (1996)

Temperature is one of the most important growth factors that govern the adaptation and speciation of plants to a wide range of environments with combinations of abiotic conditions and biotic interactions. Plants have variable complex combinations of morphological, anatomical, physiological evolved and biochemical traits that suit the species for its perpetuation in a particular habitat (Bowers, 1994). Plants can grow successfully in a enormous range of different thermal environments, extending from temperature below freezing point in arctic region to roaring temperature above 55°C of the hot desert (Table 1). Survival and growth of lower eukaryotic micro-organisms at temperature below freezing point are an example of the adaptation of plants to environmental extremes (Psenner and Sattler 1998). Some highly specialized prokaryotes can grow at temperature above 115°C, but eukaryotes appear less variable and do not normally occur above 55°C. Bioproductivity of the desert ecosystem representing arid conditions, can be comparable to that of crop-land ecosystem. A gelatinous lichen, *Collema tenax*, with its cynobacterial symbiont, is the most abundant photosynthetic form in the highly arid conditions of the desert. It can maintain photosynthetic activity over a wide range of temperature from near freezing to over 40°C, requires high light intensity for maximal photosynthesis, and as high light compensation point (the minimum light intensity at which it can break even in terms of carbon balance). The gelatinous thallus of *C. tenax* serves as a reservoir maintaining its hydrated state long enough to provide for extended photosynthesis even in the hot conditions when the light intensity is very high in the noon hours of the desert day. All of these features described for *C. tenax* are comparable with the "sun plant" and thus would be of wide taxonomic, physiological and biogeographical significance (Moore, 1998).

Maximum productivity of crops under optimal growth conditions may take place as a result of specific trait combinations that maximize yield. However, the conditions in the field are seldom appropriate for attaining the potential of a crop variety. Temperature has been rated as the most important limiting factor in crop raising. Slight variation in temperature even transiently may affect the growth and yielding ability of the crop (Singh *et.al.*, 1997). In the cropping season, daily changes in temperature from 2-10°C or even more under tropical and subtropical conditions, are quite common. Moreover, abrupt deviation from the minimum as well as maximum temperature in a locality/region, even for a short period might prove quite deleterious to growth, metabolism and yield of crops. Accordingly, temperature-stress resistance, i.e. the ability to maintain and project basic metabolic and developmental processes on exposure to temperature stress, is an important attribute of developing/characterizing crop varieties that are able to maintain yields in habitats that are of the extremes of thermaltolerance of specific crops. Improvement of thermo tolerance crops is, thus, a major target within the general objective of breeding varieties, characterized by a stable yield even in limiting environmental conditions (McKersie and Ya'acov, 1994).

The aspect of the mechanism of thermotolerance and its significance in raising crops, has attracted the attention of scientists all over the world, but traditional plant breeding approaches have, so far, met with limited success in breeding freezing tolerant and heat-stress tolerant agronomic plants. Temperature stress tolerance of the best crop varieties today is essentially the same as those varieties developed in early part of the century. The breakthrough the developing temperature-stress tolerant crop varieties is limited due to the following reasons :

- Thermotolerance is a very complex trait resulting from the optimization of several complex physiological and biochemical process at both cellular and whole plant level and at different development stages.
- Due to this complex nature and a strong genotype-environment interaction, the trait shows a low heritability and is difficult to measure, so that the classification of genotypes as tolerant or susceptible is often uncertain.
- The genetical basis of the trait is almost unknown even though a large number of genes are expected to be involved.

Current approaches for increasing crop productivity are to develop varieties that are able to maintain yield in the habitats that are at the extremes of thermal tolerance of specific crop species. Efforts are, thus, to be made for developing some of these varieties for which environmental tolerance can be expanded, so that the range of appropriate climatic conditions for crop production may be increased. An alternative to overcome the problem of breeding for environmental stress tolerance in crops, is to understand the physiological mechanism by which plant naturally tolerate environments with extreme temperature regimes. Physiological parameters generally show a higher heritability and for which accurate and reliable field and laboratory tests are available to analyse them separately (Maffat *et. Al.*, 1990; Mahajan *et. al.*,1993). This information can lead to efficient to development of new crop variety by breeding for specific adaptive traits or by genetic engineering of these traits.

Many review articles on the responses of plants to low temperature (Nishida and Murata, 1996) and high temperature stresses (Grillo and Leone, 1997; Weis and Berry, 1988) have appeared in recent years. Divergent concepts involving various models at ecological and cellular levels have suggested the role of multiple genes in temperature tolerance of plants. There is a need to reduce the gap between the conclusions drawn by the analysis of the cellular processes and by the analysis of cold and heat specific gene structure and function. Efforts in the present article have been made to address the problems of genetic diversity: (i) to pushing back the limits imposed by the temperature stress, and (ii) to focus on unified theory of temperature tolerance in the existing plant species and the future crop varieties.

INDICES OF TEMPERATURE RESPONSES OF PLANTS

There exists a lot of variability in the responses of crop species to low as well as high temperature stresses. In broader sense, plants have been categorized as sensitive, facultatively tolerant and tolerant to temperature stress (Nilsen and Orcutt, 1996). Sensitive species are those that are inhibited by relatively low temperatures and lack the capacity to adjust their heat sensitive response to increasing ambient temperature. Low temperature sensitive species will be injured at tissue temperature between 30 and 40°C. Facultative heat tolerant species can adjust the cytoplasmic heat tolerance in response (above 50°C) is not commonly observed in the crop plants, except the seeds, which have evolved the unique mechanism of resumption of growth (germination) from highly dehydrated state attained during maturity. Conventionally, range of temperature variability (minimum, optimum, maximum) in the cropping season, has been defined as a the cardinal points, which are used to get an appropriate idea about the temperature requirements and the level of the temperature regimes at which it becomes stress for the growth and yield of crops (Table 2).

TABLE 2

Cardinal points (°C) for crops grown in cold (*rabi*) and warm (*kharif*) seasons

S. No.	Response	Cool season	Warm season
1	Minimum temperature below which growth is inhibited	0-5	15-18
2	Optimal temperature which supports maximum growth	25-30	30-37
3	Maximum temperature above which growth cases	30-37	44-50

Source: Dhaliwal and Kler (1995)

LOW TEMPERATURE STRESS

Low temperature is critical to plant in that it is an important determinant in the normal distribution of natural plant species, and defines the distribution and growth of important agricultural crops, many of which are grown at or near the temperature boundary-limits of these genetically determined survival abilities.

Low temperature occurs seasonally over approximately 90 per cent of earth dry land. Chilling stress is usually limited to plants native to or growing in tropical and subtropical regions. The capability to withstand freezing is one of the physiological traits for geographical distribution and productivity of native and cultivated plant species grown in cold climate/cold season. Plants adapted to low temperature stresses have evolved growth habits, morphological features and efficient biochemical and physiological strategies that allow them to perpetuate and grow in such hostile environments (Veziana *et.al.*, 1996). The acclimation processes are similar in plants subjected to chilling and freezing stress, except that in the latter the responses are accentuated and ice formation is involved. Generally, intracellular ice formation is injurious to most plants (Nilsen and Orcutt, 1996).

Some of the plant species exhibiting chilling resistance and freezing susceptibility do acquire a high degree of resistance to chilling when exposed to non-injurious temperature. Development of low temperature tolerance in plants after being exposed to low non-freezing temperature tolerance in plant after being exposed to low non-freezing temperature for several days, is the adaptive process known as acclimation (Guy, 1990).

MECHANISM OF ACCLIMATION AND ADAPTATION

Cold signal transduction. The primary events occurring during low temperature acclimation related with cold tolerance in plants are designated as signal transduction. Cold sensor probably resides in the plasma membrane and acts in message transfer through changes in its fluidity under different temperature regimes (Uemura *et.al.*, 1995). Vigh *et.al.* (1993) have demonstrated that temperature induced changes in membrane fluidity trigger the expression of a mRNA encoding for a desaturase enzymes which catalyzes the synthesis of unsaturated fatty acid forming the part of the fluid membrane. This has been confirmed by Maresca and Cossins (1993) that a *Synechocystis* the membrane acts as a sensor for low temperature and affects nuclear transcriptional activity. Abscisic acid (ABA) is a primary hormone that mediates plant responses to abiotic stresses, like chilling and freezing, drought, desiccation, salinity and anaeroblosis. Since ABA can increase frost tolerance in non-acclimated plants, it can act as a signal transducer. Presence of ABA-responsive elements in the 5-upstream sequences of many genes, whose transcription is unregulated during cold acclimation and upregulation of transcription of cold induced gene encoding for protein kinase, strengthens the role of ABA as single transducer (Holappa and Walker-Simmons, 1995). One of the ways through which ABA regulates many genes, is the cellular concentration of calcium and its effect on cellular transduction (Pei *et.al.*, 1998).

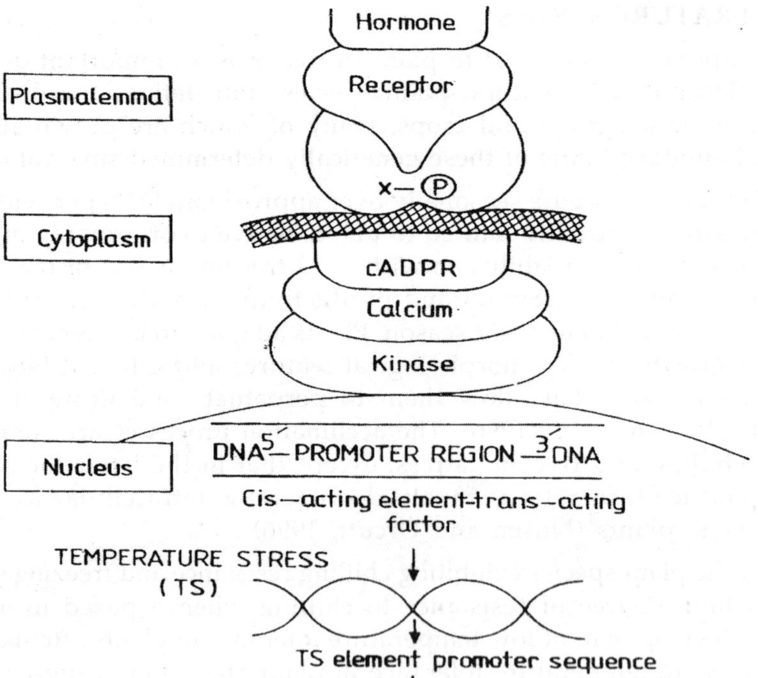

Fig. 1 : Molecular model of temperature stress transcription factor (TSF) activation and regulation of temperature stress response in plants

ABA-mediated increases in Ca^{2+} levels in the guard cell, leads to stomatal closure (McAinsh *et.al.*, 1990) and Ca^{2+} can also regulate the expression of ABA-responsive gene in maize protoplast (Sheen, 1996). Wu *et.al.* (1997) demonstrated that cyclic ADP ribose (cADPR) is a likely *in vivo* intermediate of ABA signal transduction that exerts its efforts by way of intracellular calcium release, and thus, triggers stress responses in plants. Bioassay experiments showed that the amounts of cADPR in the Arabindopsis thaliana tissues increased in response to ABA treatment and before ABA induced gene expression. Wu *et.al.* (1997), thus, proposed that cADPR is a second messenger responsible for initiating a cascade of Ca^{2+} increases, and subsequently Ca^{2+} dependent phosphorylation during ABA signal transduction. cADP-ribose looks like a ubiquitous messenger across the plant and animal kingdom. Information about the single molecular mechanism involving ABA, cADPR and Ca^{2+}, opens the way to produce crops that resist stress better perhaps by engineering them to respond more quickly and efficiently under stress conditions. However, in order to design methods to engineer plants to resist temperature stress *vis-à-vis* thermotolerance, it is essential to know about the signal to be transmitted. For instance, besides cADPR, intracellular Ca^{2+} levels respond to many other signals, like cAMP, IP_3, calmodulin. There are multiple mechanisms for the release of calcium. *In vivo,* these multiple mechanisms may cause calcium to be released from different storage sites and different patterns, generating a calcium signature. This is how the cell can keep its signals straight enough though the same ion calcium is involved in many different signal pathways. Clarification on this point will be useful in understanding the mechanism of the regulation of stress tolerance in plant (Fig.1)

Cold induced gene expression. Process of cold acclimation involves the adaptation to low temperature and development of chilling and freezing tolerance, both of which are associated with biochemical and physiological changes and alterations in the cold regulated (CoR) gene expression (Guy, 1990). Genetic analysis has indicated that multiple genes are involved in cold acclimation in plants (Hajela *et.al.*, 1990). Constitutive expression of CoR gene, i.e. CoR15a, which encodes the chloroplast targeted polypeptide (CoR 15 am) in transgenic *Arabidopsis* plants, improves the freezing tolerance in chloroplast frozen *in situ* and chloroplast frozen *in vitro*. Several CoR genes i.e. CoR78, CoR 66 are co-ordinately stimulated along with CoR 15a in response to low temperature, which suggests that CoR 15a might act in a concert with other CoR genes to enhance tolerance to freezing in plants. If so, expression of the entire battery of CoR genes would have a greater effect on freezing tolerance than the expression of CoR gene alone.

Trans-acting factors (commonly proteins) reorganize and bind to the *cis*-acting elements located within promoter region of the eukaryotic gene; these factors control synthesis of mRNA by RNA polymerase II. Regulatory proteins that are synthesized in the cytosol and targeted to the nucleus, bind to *cis*-acting elements in the 5′-upstream regions of genes encoding for proteins with various non-regulatory functions. The binding of these factors together with binding of other less specific nuclear proteins, modify the upstream region of the targeted genes and initiate a transcriptional wave. The cascade of events is termed as transcriptional regulation.

Cold-induced synthesis of trans-acting factor has suggested that cold modulation of gene expression through a cascade of molecular events, and that *trans*-acting factors play an important role in initiating a transcription wave that results in such cascade. A *cis*-acting element that confers cold inductability has been isolated and sequenced in *A thalians* contains four regions harboring *cis*-acting elements that interact to control transcription under various stress conditions including low temperature stress.

Wang *et.al.,* (1995) demonstrated that once cold signal is transduced, specific regulating elements act to modulate transcription. *Trans*-acting factors that bind to these *cis*-acting elements have been isolated from the pool of nuclear proteins. Also, cold specific *trans*-acting factors have been isolated and characterized in *A. thaliana*. This has implication in characterizing frost tolerance in oilseed crop *Brassica*, a close phylogenetic relative of *A. thaliana*, where cold tolerance is an important agronomic trait.

The regulatory element is present in the promoters of multiple CoR genes including CoR78, Cor 66 and CoR15a. *Cis*-acting drought responsive element, CRT/DRE, is the DNA regulatory element that stimulates transcription and CBF, *trans*-acting factor (regulatory gene factor) that binds to the CRT/DRE sequence, has been shown as the putative transcriptional activator of CoR genes in response to both low temperature and moisture deficit (Yamaguchi-Schinozaki and Schinozaki, 1994). Cold acclimation is linked with the induction of CoR genes mediated by the drought responsive (CRT/DRE) DNA regulatory elements.

Jaglo-Ottosen *et.al.,* (1998) reported that increased expression of *Arabidopsis* CBF_1 a transcriptional activator that binds to the CRT/DRE sequence, induced CoR gene expression and increased the freezing tolerance of non-acclimated plants, suggesting that CBF_1 is the regulator of the cold acclimation response, controlling the level of CoR gene expression, which in turn promotes freezing tolerance. It is likely that CRT/DRE containing gene and CBF, as its transcription activator, are not limited to *Arabidopsis* (Jiang *et.al.,* 1996) and thus may provide a way to improve low temperature tolerance in the crop plants. Delineation on the involvement of CRT/DRE containing CoR genes in bringing about the full array of biochemical and physiological changes will enhance our understanding about cold acclimation *vis-à-vis* improvement in low temperature tolerance in crops. *In situ* plants co-ordinate the expression of their cold regulated genes, turning them all together as the temperature drops. Known cold regulated genes possibly carrying the same regulatory sequence, give an indication that they are all turned on by the same transcription factor (master switch). It might be possible to take advantage of this synchrony in understanding the mechanism of cold tolerance in crop plants (Murata *et.al.,* 1992).

***Cold regulated protein synthesis*.** Cold acclimation is related with the production of CoR proteins and enhancement of the freezing tolerance (Hughes and Dunn, 1996). Some CoR genes code for proteins that seem to protect plant cell membrane against the disruption that results when freezing or other dehydration conditions deplete the water molecular normally found in the space between the membrane and the rough cell wall. Cold related proteins in wheat build up in this gap and seem to help keep the membrane intact (Pearce and Ashworth, 1992). Cold acclimation proteins may act as anti-freeze proteins (AFPs), which could increase super cooling, reduce the amount of freezable water at a given temperature and decrease the rate of ice spread in the tissues (Antikainen and Grifith, 1997; Worral *et.al.*, 1998).

Significant levels of cold induced antifreeze proteins were present only in the apoplast of freezing tolerant and closely related group of plants, i.e. rye, wheat and barley. The results indicate that the accumulation of antifreeze proteins is a specific response that may be important in freezing tolerance of these plants rather than a general response of all plants to low temperature stress (Graham and Patterson, 1982). Accumulation of dehydrin-like proteins of 65, 60 and 14 KDa during cold acclimation and their association with cold hardiness, has been reported in blue-berry cultivars, suggesting that changes in dehydrin expression are more closely associated with cold hardiness (Arora *et.al.*, 1997). The cDNA (F cor1) is upregulated by low temperature in cold acclimated strawberry, more so in hardy plants. The F cor 1 transcript encoded protein has a compositional bias for leucine, isoleucine, glycine, proline and serine in cold acclimated strawberry plants. This protein shows homology with the protein encoded by blt 101, a low temperature responsive gene from barley, and ESI 3, a gene induced by salt stress in *Lophopyrum* (NDong *et.al.*, 1997). It is possible that F cor 1 may serve as a suitable marker to select for low temperature tolerance in related species of the rosaceae family in particular and other species in general.

Uncoupling proteins (UPCs) which dissipate the proton electrochemical potential gradient across the inner mitochondrial membrane are implicated in the heat production in plants. Laloi *et.al.*, (1997) have identified cDNA from potatoes (*Solanum tuberosum*) that encodes a plant coupling protein, St. UCP, which is induced by cold treatment and this may be involved in thermoregulation in plants. St. UCP is expressed in fully expanded leaves, stems, fruit, stolon, roots, flower as well as in tubers. Therefore, St. UCP is ubiquitously expressed and may be involved in thermoregulation in plants regardless of development stage and plant parts. Cold induced expression of uncoupling protein (St. UCP) might contribute to these phenomena in combination with the alternative oxidase to be involved in heat production vis-à-vis thermotolerance in plants.

Membrane structure and function. Information on the mechanism of freezing injury in plants has led to the generally accepted concept that irreversible damage to freezing induced dehydration to the plasmalemma as well as vacuolar membranes, is the primary cause of freezing injury (Murai and Yoshida, 1998). In mungbean of subtropical origin low temperature caused dysfunction of the vacuolar membrane, particularly the impairment of protein translocation system (Matsuura-Enlo *et.al.*, 1992).

The damage to plasma membrane as a consequence of freeze dehydration is closely associated with particular changes in the membrane structure, resulting into loss in the membrane permeability (Kodama *et.al.*, 1995). Thus, a particular molecular mechanism which stabilizes membrane bilayers against freeze induced dehydration may play an important role in the cold acclimation in plants. Kojma *et.al.*, (1998) have suggested that in Adjuki bean (*Vigna angularis*), a chilling sensitive plant, cold acclimation increases the relative amount of phospholipids and decreases sterols in the membrane phospholipids of acclimated plants which have higher percentage of unsaturated fatty acid chain than those of control. Lipid unsaturation is one the main factors maintaining the membrane fluidity at low temperature, consistent with low temperature stress resistance in plants (Nishida and Murata, 1996).

Low temperature favoured the accumulation of two unsaturated fatty acids, i.e. linoleic acid. At low temperature the tails of saturated fatty acids are packed closely binding the movement of membrane protein (solidification), so damaging the cell. However, unsaturated fatty acids make the membrane more fluid, because the double bond kink modifies the tail of unsaturated fatty acid so that it packs less easily in the centre of the membrane.

HIGH TEMPERATURE (HEAT) STRESS

Every organism expresses thermotolerance for its survival at higher temperature (Singla *et.al.*, 1996). Exposure to heat (heat shock) increases the basic level of its performance. This acquisition (induction) of a higher level of thermotolerance protects the cells, tissues, organs and the organisms from a subsequent, otherwise, lethal high temperature (Lie *et.al.*, 1995). Biochemical and physiological changes elicited by sudden and transient increase in temperature is termed as heat stock (HS) response, which will include the changes, in the physiological, biochemical and molecular processes.

Heat-stress Threshold

A threshold temperature refers to a value of daily mean temperature at which a detectable reduction in growth begins. Upper and lower developmental threshold temperatures have been determined for many plant species through controlled laboratory and field experiments. A lower developmental threshold or a base temperature is one below which plant growth and development stop. Similarly, an upper developmental threshold is the temperature above which growth and development cease. Knowledge of lower threshold temperatures is important in physiological research as well as for crop production. Base threshold temperatures vary with plant species, but for cool season crops 0 °C is often the best-predicted base temperature (Miller *et.al.*, 2001). However, determining a consistent upper threshold temperature is difficult because the plant behavior may differ depending on other environmental conditions (Miller *et.al.*, 2001).Upper threshold temperatures for some major crop species are diplayed in Table 2. Brief exposure of plants to high temperatures during seed filling can accelerate senescence, diminish seed set and seed weight, and reduce yield (Siddique *et. al.*, 1999). This is because under such conditions plants tend to divert resources to cope with the heat stress and thus limited photosynthates would be available for reproductive development. Another effect of temperature stress in many plant species is induced sterility when temperature is imposed immediately before or during anthesis Pulse legumes are particularly sensitive to heat stress at the bloom stage; only a few days of exposure to high temperatures (30-35 0c) can cause heavy yield losses through flower drop or pod abortion (Siddique *et.al.*, 1999). Thus, it is highly desirable to appraise threshold temperatures of new cultivars to prevent damages by unfavorable temperatures during the plant ontogeny.

TABLE 2

Threshold high temperatures for some crop plants

Crop plant	Threshold Temper(°C)	Growth stages	References
Wheat	26	Post anthesis	Stone and Nicol´as (1994)
Corn	38	Grain filling	Thompson (1986)
Cotton	45	Reproductive	Rehman *et.al.,* (2004)
Pearl millet	35	Seedling	Ashraf and Hafeez (2004)
Tomato	30	Emergence	Camejo *et.al.,* (2005)
Brassica	29	Flowering	Morrison and Stewart (2002)
Cool season pulses	25	Flowering	Siddique *et.al.,* (1999)
Groundnut	34	Pollen productions	Vara Prasad *et.al.,*(2000)
Cowpea	41	Flowering	Patel and Hall (1990)
Rice	34	Grain yield	Morita *et.al.,* (2004)

TEMPERATURE STRESS ACCLIMATION AND ADAPTATION

Heat shock proteins (HSPs). Synthesis and accumulation of specific proteins are ascertained during a rapid temperature stress, and these proteins are designated as HSPs. Increased production of HSPs occurs when plants experience either abrupt increase in temperature (Nakamoto and Hiyama, 1999). In higher plants, HSPs are usually induced under heat shock at any stage of development and major HSPs are highly homologous among distinct organisms (Prasinos *et.al.,* 2005). HSP-triggered thermotolerance is attributed to the observations that (a) their induction coincides with the organism under stress, (b) their biosynthesis is extremely fast and intensive, and (c) they are induced in a wide variety of cells and organisms. Three classes of proteins, as distinguished by molecular weight, account for most HSPs, *viz.,* HSP90, HSP70 and low molecular weight proteins of 15-30 kDa. The special importance of small HSPs in plants is suggested by their unusual abundance and diversity. The proportions of these three classes differ among plant species. HSP70 and HSP90 mRNAs can increase ten- fold, while low molecular weight (LMW) HSPs can increase as much as 200-fold. Other proteins, such as 110 kDa polypeptides and ubiquitin, though less important, are also considered to be HSPs (Feussner *et.al.,* 1997). The gene for a nuclear-encoded HSP, *Hsa32,* encoding a 32 kDa protein, has been cloned in tomato (Liu *et.al.,* 2006). When maize, wheat and rye seedlings were subjected to heat shocks (42°C), whereas five mitochondrial LMW-HSPs (28, 23, 22, 20 and 19kDa) were expressed in maize, only one (20 kDa) was expressed in wheat and rye, suggesting the reason for higher heat tolerance in maize than in wheat and rye (Korotaeva et aI., 2001). Presence of HSPs can prevent denaturation of other proteins caused by high temperature. The conformational dynamism and aggregate state of small HSPs may be crucial for their functions in thermoprotection of plant cells from detrimental effects of heat stress (Iba, 2002).

In response to high temperatures, specific HSPs have been identified in different plant species. For example, HSP68, which is localized in mitochondria and normally constitutively expressed, was determined to have increased expression under heat stress in cells of potato, maize, tomato, soybean and barley (Neumann *et.al.,* 1993). Fast accumulation of HSPs in sensitive organs/tissues can play an important role in protection of metabolic apartment of the cell, thereby acting as a key factor for plant's adaptation to, and survival under, stress. Many studies assert that HSPs are molecular chaperones insuring the native configuration and functionality of cell proteins under temperature stress. There is considerable evidence that acquisition of thermotolerance is directly related to the synthesis and accumulation of HSPs (Bowen *et.al.,* 2002). Recently, dual role of LMW HSP21 in tomato has been described as protecting PSII from oxidative damage and involvement in fruit color change during storage at low temperatures (Neta-Sharir et al., 2005). In many plant species, thermo tolerance of cells and tissues after a heat stress is pretty much dependent upon induction of HSP70, though HSP101 has also been shown to be essential (SchOffl *et.al.,* 1999). One hypothesis is that HSP70 participates in ATP-dependent protein unfolding and it prevents protein denaturation during heat stress (Iba, 2002). Temperature sensitivity was associated with reduced capacity of bentgrass variants to accumulate chloroplastic HSPs (Wang and Luthe, 2003). LMW-HSPs may play structural roles in maintaining cell membrane integrity. Localization of LMW-HSPs in chloroplast membranes further suggested that these proteins protect the PSII from adverse effects of temperature stress and play role in photosynthetic electron transport (Barua *et.al.,* 2003). Exposure of cells or tissues of various organisms to elevated temperature induces the synthesis of a family of specific proteins called heat shock proteins (HSPs). HSPs can be induced in bacteria, plants, animals and fungi and fall into different classes: 70 Kd HSPs and a low molecular weight 14-30 Kd class. Thermal shock can induce HSP 29, 27 and 70 in all organisms. The principal heat shock proteins have many roles to play relating to high temperature tolerance (Lindquist, 1986).

- HSPs that protect metabolic processes and the newly made proteins from misfolding belong to five conserved classes: HSP 100, HSP90, HSP70, HSP60 and the small heat shock protein (5 HSPs). 5 HSPs have a wide range of cellular functions, i.e. (i) endowing cells with thermotolerance *in vivo* and (ii) being able to act as molecular chaperones *in vitro*. Chaperone heat shock proteins arise to refold proteins that have been desaturated by excessive heat, and 5 HSPs do this by forming stable complexes with folding intermediate of their protein substrates. There is little information available about these structures and the mechanism by which substrates are protected from thermal denaturation by 5 HSPs (Kim *et.al.,* 1998).
- Protease type HSPs degrade proteins that have been permanently damaged.
- Constitutive HSPs may be able to protect some aspects of cell metabolism from high heat damage. These HSPs are thought to function as molecular chaperones, which serve to facilitate protein folding and transport of proteins across membrane of proteases that facilitate the recycling of amino

acids from damaged proteins into new functional proteins (Mariama *et.al.*, 1997).

Regulation of HSPs gene expression. Induction of HSPs following heat shock is dependent both on transcription and translation. Regulation of transcription of HSPs is through a heat shock element (HSE). HSEs are present on the DNA upstream from all HSP genes and contains small consensus sequences (sections of nucleotide sequence that are identical among all HSEs from all organisms). Induction of heat shock gene transcription occurs when a heat shock factor (HSF, a transcription regulatory protein) binds to the HSE upstream from HSP gene (Bienz and Pellham, 1987). The interaction between HSP, HSF and HSE, regulates the short term induction of HSP gene transcription (Fig.1). Thus, the response of HSP gene transcription to heat shock is a fine - tuned process that opens up many possibilities for genetic engineering and induction of heat stress tolerance in plants.

Enzymes. High temperature causes denaturation of the protein, coinciding with drop in the function of the enzyme due to the loss of tertiary structure of the protein at high temperature as a consequence of breaking up of the weak molecular bonds that keep a polypeptide chain folded appreciably. Those proteins that have disulfide bonds in tertiary structure are relatively resistant to denaturation, and those proteins that have complex structure, based on the diversity of relatively weak molecular bonds as in most of the enzymes, are sensitive to heat-induced denaturation. Thermal stability of enzymes can vary to some degree amongst species that grow on different environments. Differential thermal stability of isoenzymes has been reported as one of the parameters associated with high temperature tolerance. Hydroxypyruvate reductase and glutathione reductase are two thermostable enzymes which can play important role in protecting plants from heat stress (Comis, 1992).

Physiological Responses

Plant water status is the most important variable under changing ambient temperatures (Mazorra *et.al.*, 2002). In general, plants tend to maintain stable tissue water status regardless of temperature when moisture is ample; however, high temperatures severely impair this tendency when water is limiting (Machado and Paulsen, 2001). Under field conditions, high temperature stress is frequently associated with reduced water availability (Simoes-Araujo *et.al.*, 2003; Anon *et.al.*, 2004). In sugarcane and tamato, leaf water potential and its components were changed upon exposure to heat stress even though the soil water supply and relative humidity conditions were optimal, implying an effect of temperature stress on root hydraulic conductance (Wahid and Close, 2007; Morales *et.al.*, 2003). In general, during daytime enhanced transpiration induces water deficiency in plants, causing a decrease in water potential and leading to perturbation of many physiological processes (Tsukaguchi *et.al.*, 2003).

A key adaptive mechanism in many plants grown under extreme temperatures, is accumulation of certain organic compounds of low molecular mass, generally referred to as compatible osmolytes (Sakamoto and Murata, 2002). Under stress, different plant species may accumulate a variety of osmolytes such as sugars and sugar alcohols, proline, ammonium compounds, and tertiary sulphonium compounds (Sairam and Tyagi, 2004).

Glycinebetaine (GB), an amphoteric quaternary amine, plays an important role as a compatible solute in plants under various stresses, such high temperature (Sakamoto and Murata, 2002). Capacity to synthesize GB under stress conditions differs from species to species such as maize and sugarcane (Ashraf and Foolad, 2007; Quan *et.al.*, 2004; Wahid and Close, 2007) due to desiccating conditions of water deficit or high temperature. In contrast, plant species such as rice *(Oryza sativa)*, mustard *(Brassica* sp.), *Arabidopsis (Arabidopsis thaliana)* and tobacco *(Nicotiana tabacum)* naturally do not produce GB under stress conditions. However, genetic engineering has allowed the introduction of GB-biosynthetic pathways into GB-deficit species (Sakamoto and Murata, 2002; Quan *et.al.*, 2004). GB, proline is also known to occur widely in higher plants and normally accumulates in large quantities in response to environmental stresses (Kavi Kishore *et.al.*, 2005). The functional significance of accumulation of compatible solutes, proline or GB synthesis may buffer cellular redox potential under temperature and other environmental stresses in sugarcane (Wahid and Close, 2007). Under high temperatures, fruit set in tomato plants failed due to the disruption of sugar metabolism and proline transport during the narrow window of male reproductive development (Sato *et.al.*, 2006). Because, The significant roles of osmolytes in response to environmenal stresses in plants, crop stress (e.g., heat) tolerance might be enhanced by increased accumulation of compatible solutes through genetic engineering (GE) approaches (Ashraf and Foolad, 2007).

Photosynthesis

Alterations in various photosynthetic attributes under temperature stress are good indicators of thermo tolerance of the plant as they show correlations with growth. Any constraint in photosynthesis can limit plant growth at high temperatures. Photochemical reactions in thylakoid lamellae and carbon metabolism in the stroma of chloroplast have been suggested as the primary sites of injury at high temperatures (Wise *et.al.*, 2004). Chlorophyll fluorescence, the ratio of variable fluorescence to maximum fluorescence (F_v/F_m), and the base fluorescence (F_0) are physiological parameters that have been shown to correlate with temperature tolerance (Yamada *et.al.*, 1996). Increasing leaf temperatures and photosynthetic photon flux density influence thermotolerance adjustments of PSII, indicating their potential to optimise photosynthesis under varying environmental conditions as long as the upper thermal limits do not exceed (Marchand *et.al.*, 2005). Under high temperatures, degradation of chlorophyll *a* and *b* was more pronounced in developed compared to developing leaves (Wahid and Ghazanfar, 2006). PSII is highly thermolabile, and its activity is greatly reduced or even partially stopped under high temperatures (Camejo *et.al.*, 2005), temperature stress may lead to the dissociation of oxygen evolving complex (OEC), resulting in an imbalance between the electron flow from OEC toward the acceptor side of PSII in the direction of PSI reaction center (Fig. 2) (De Ronde *et.al.*, 2004). Heat stress causes dissociation of a manganese (Mn)-stabilizing 33-kDa protein at PSII reaction center complex followed by the release of Mn atoms (Yamane *et.al.*, 1998).

In wheat, high temperatures and excessive light dam- aged different sites of PSII, which implied different pathways for the recovery of its functional activity (Sharkova, 2001). In barley, temperature pulses abruptly damaged the PSII units and caused loss of their capacity of oxygen evolution leading to a restricted electron transport, which was totally abolished after four hours (Toth *et.al.*, 2005).

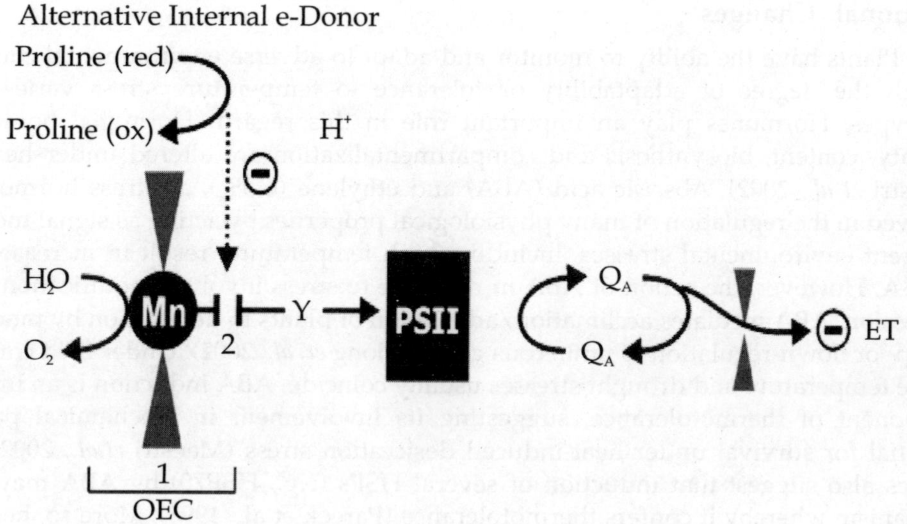

Fig. 2 : Temperature induced inhibition of oxygen evolution and PSII activity

High temperature influences the photosynthetic capacity of C3 plants more strongly than in C4 plants. It alters the energy distribution and changes the activities of carbon metabolism enzymes, particularly the rubisco, thereby altering the rate of RuBP regeneration by the disruption of electron transport and inactivation of the oxygen evolving enzymes of PSII (Salvucci and Crafts-Brandner, 2004b). Heat shock reduces the amount of photosynthetic pigments (Todorov *et.al.*, 2003), soluble proteins, rubisco binding proteins (REP) and large-(LS) and small subunits *(SS)* of rubisco in darkness but increases them in light, indicating their roles as chaperones and HSPs (Kepova *et.al.*, 2005).

Assimilate Partitioning

Under low to moderate temperature stress, a reduction in source and sink activities may occur leading to severe reductions in growth, economic yield and harvest index. Assimilate partitioning, taking place via apoplastic and symplastic pathways under high temperatures, has significant effects on transport and transfer processes in plants (Taiz and Zeiger, 2006). However, considerable genotypic variation exists in crop plants for assimilate partitioning, as in wheat genotypes (Yang *et. al.*, 2002). To elucidate causal agents of reduced grain filling in wheat under high temperatures, Wardlaw (1974) examined three main components of the plant system including source (flag leaf blade), sink (ear), and transport pathway (peduncle). It was determined that photosynthesis had a broad temperature optimum from 20 to 30°C, however it declined rapidly at temperatures above 30°C. The rate of 14c assimilate movement out of the flag leaf (phloem loading), was optimum around 30°C, however, the rate of movement through the stem was independent of temperature from 1 to 50 °C. Wheat, temperature effects on translocation result indirectly from temperature effects on source and sink activities. From such results, increased mobilization efficiency of reserves from leaves, stem or other plant parts has been suggested as a potential strategy to improve grain filling and yield in wheat under heat stress.

Hormonal Changes

Plants have the ability to monitor and adapt to adverse environmental conditions, though the degree of adaptability or tolerance to temperature stress varies among genotypes. Hormones play an important role in this regard. Hormonal homeostasis, stability, content, biosynthesis and compartmentalization are altered under heat stress (Maestri *et al.*, 2002). Abscisic acid (ABA) and ethylene (C_2H_4), as stress hormones, are involved in the regulation of many physiological properties by acting as signal molecules. Different environmental stresses, including high temperature, result in increased levels of ABA. However, the action of ABA in response to stress involves modification of gene expression. ABA mediates acclimation/adaptation of plants to desiccation by modulating the up- or down-regulation of numerous genes (Xiong *et. al.*, 2002). Under field conditions, where temperature and drought stresses usually coincide, ABA induction is an important component of thermotolerance, suggesting its involvement in biochemical pathways essential for survival under heat-induced desiccation stress (Maestri *et.al.*, 2002). Other studies also suggest that induction of several HSPs (e.g., HSP70) by ABA may be one mechanism whereby it confers thermotolerance (Pareek et aI., 1998). More so, heat shock transcription factor 3 acts synergistically with chimeric genes with a small HSP promoter, which is ABA inducible (Bannu EFEGLU, 2009).

A gaseous hormone, ethylene regulates almost all growth and developmental processes in plants, ranging from seed germination to flowering and fruiting as well as tolerance to environmental stresses. Ethylene has nearly full biological activity at 1 fLLL -1, corresponding to 6.5 x 10^{-9} M at 25°C. However, the levels of ethylene or the enzymes involved in ethylene biosynthesis vary at different time intervals during the day. For instance, the endogenous concentration of l-amino-cyclopropane-l-carboxylic acid (ACC), a precursoı of ethylene biosynthesis, measured at predawn and at maximum solar radiation during a summer drought showed a sharp distinction between the two times, which was positively correlated with the intensity of incident solar radiations (Munne-Bosch *et.al.*, 2002).

Temperature stress changes ethylene production differently in different plant species (Arshad and Frankenberger, 2002). For example, while ethylene production in wheat leaves was inhibited slightly at 35°C and severely at 40°C, in soybean ethylene production in hypocotyls increased by increasing temperature up to 40°C and it showed inhibition at 45 °C. Temperatures up to 35°C have been shown to increase ethylene production and ripening of propylene-treated kiwifruit, but temperature above 35 °C inhibits ripening by inhibiting ethylene production, although respiration continues until the tissue disintegration (Antunes and Sfakiotakis, 2000). Ethylene may overcome the inhibitory effect of high temperature on thermosensitive lettuce seed due to increased a-mannanase activity, which helps weakening of the endosperm and facilitates germination (Nascimento *et. al.*, 2004). The effect of pre-harvest temperature on ripening characteristics of shaded and sun-exposed apple fruits indicated that the former treatment produced up to 90% less ethylene than the latter (Klein *et. al.*, 2001). In maize, ethylene production was highest at the top ear and lowest at the middle ear, suggesting that ethylene plays an important role in assimilate partitioning to grain filling (Fenglu *et. al.*, 1997). Among other hormones, salicylic acid (SA) has been suggested to be involved in heat-stress responses elicited by plants. SA is an important component of signaling pathways in response to systemic acquired resistance (SAR) and the hypersensitive response (HR) (Kawano *et.al.*, 1998). SA stabilizes the trimers of heat shock transcription factors and aids them bind heat shock elements to the promoter of heat shock related genes. Longterm thermo tolerance can be induced by SA, in which both Ca^2+ homeostasis and antioxidant systems are thought to be involved (Wang and Li, 2006b). Sulphosalicylic acid (SSA), a derivative of SA, treatment can effectively remove H_2O_2 and increase heat tolerance. In this regard, catalase (CAT) plays a key role in removing H_2O_2 in cucumber *(Cucumus sativus)* seedlings treated with SSA under temperature stress. Thermotolerance of plants also can be enhanced by spraying leaves with acetyl-SA by enhanced xanthophylls de-epoxidation and content of ascorbate, antioxidants and a-tocopherol in leaves (Llusia *et. al.*, 2005; Wang and Li, 2006a).

The effects of gibberellins and cytokinins on high temperature tolerance are opposite to that of ABA. An inherently heat-tolerant dwarf mutant of barley impaired in the synthesis of gibberellins was repaired by application of gibberellic acid, whereas application of triazole paclobutrazol, a gibberellin antagonist, conferred heat tolerance (Vettakkorumakankav *et. al.*, 1999). In a dwarf wheat variety, high temperature-induced decrease in cytokinin content was found to be responsible for reduced kernel filling and its dry weight (Banowetz *et. al.*, 1999). Another class of hormones, brassinosteroids have recently been shown to confer thermotolerance to tomato and oilseed rape *(Brassica napus)*, but not to cereals (Dhaubhadel *et.al.*, 1999). The optimum temperature for photosynthesis is quite variable among habitats or seasons for one species. Temperate plant species commonly have optimum temperature for photosynthesis between 20 and 30°C and desert species may have optimal temperature as high as 45°C. The temperature optimum for photosynthesis for *Atriplex sabulosa* and *Tidestromia oblongifolia* was 25° and 45°C, respectively (Berry and Raison, 1981). The main feature of the thermal tolerance of photosynthesis in both the species was photosystem II activity. Quantum yield of photosystem II driven-electron transport decreased at a temperature that corresponded with inhibition of photosynthesis in both species. Saturation of the thylakoid membrane at higher temperature in the sensitive species causes solidification, consistent with loss in the structure of PS II light harvesting complexes and inhibition of PS II electron transport (Pearcy *et. al.*, 1977). In photosynthetic cynobacteria, i.e. Synechocystis PCC 6803, desaturation of fatty acid is altered through transformation of fatty acid by desaturation, and unsaturation of membrane lipid is found to stabilize photosynthesis against heat activation (Vigh *et.al.*, 1993).

PS II protein complexes may be stabilized by low molecular weight HSPs under heat stress conditions (Coleman and McConnaughay, 1994). Maintenance of photosynthesis in plants is a critical aspect of thermotolerance and photosynthesis is particularly sensitive to temperature stress (Weis and Berry, 1998). Certain HSPs (HSP 21 and HSP 24) are translocated into the chloroplast and correlate with increasing thermotolerance of photosynthesis. These HSPs are expected to protect PS II activity during heat stress, being the part of thylakoid membrane (Osteryoung and Veirling, 1994). Differential expression of these HSPs may result in differential thermal tolerance of PS II. Threshold for denaturation of PS n thylakoid structure in *Aegilops* and *Triticum durum* wheat genotypes was at 2511C and ranged between 35°-45°C. *Aegilops* genotype proved sensitive and *Triticum durum* genotypes resistant to high temperature stress. Cham I durum wheat was robust to both chilling stress and heat stress (Rekika *et.al.*, 1997).

Osmoprotectants. Desiccation and dehydration due to heat stress inflict injury to the plants through destabilization of the plasma membrane. Natural osmolytes, especially soluble sugars are widely used as stabilizers of plant oligomeric proteins, protecting them against a variety of adverse conditions. Soluble sugars cause acclimation of photosynthesis to high temperature in desert plants (Seemann *et.al.*, 19R6). Non-red using disaccharide, trehalose, has been used as a stabilizing agent on membrane phospholipids, proteins and cell organelles. Higher concentration of trehalose in desiccation tolerant resurrection plants (Bianchi *el.al.*, 1993) has suggested that this molecule may be involved in the desiccation tolerance of plants. This disaccharide from natural plant source has been used as stabilizer of plant oligomeric enzyme phosphoenolpyruvate-carboxylase (PEPC) which is known to be sensitive under low and high temperature stress (Salilhas el al., 1997). Lower concentration (60 mM) of trehalose fully protects the activity of PEP carboxylase from C_4 grass, *Cynodon dactylon*, against cold (O°C) activation and heat (37°C) inactivation. At temperature higher than 45°C, enzyme was better protected by trehalose (1.25 M) than glycerol or phosphate. Trehalose and oligosaccharides-stachyose and raffinose-stabilize the enzyme and proteins through preferential exclusion from protein domain, resulting in a mutual repulsion between enzyme and sugar molecule.

Proline accumulation in plant tissues under dehydration and desiccation has indicated that it is the most potent osmoprotectant that plays a role in counteracting the effect of osmotic and heat stresses. Overproduction of proline has been correlated with osmotic stress tolerance in transgenic tobacco plants (Kavi Kishore *et.al.*, 1995). It is suggested that genetically engineered crop plants that overproduce proline under temperature stress might, thus, acquire osmotolerance, i.e., the ability to tolerate environmental stresses such as drought and heat stress (Yoshiba *et.al.*, 1997).

MOLECULAR RESPONSES

Oxidative Stress and Antioxidants

In addition to tissue dehydration, temperature stress may induce oxidative stress. (Liu and Huang, 2000). AOS cause the autocatalytic peroxidation of membrane lipids and pigments thus leading to the loss of membrane semi-permeability and modifying its functions (Xu *et.al.*, 2006). Superoxide radical is regularly synthesized in the chloroplast and mitochondrion and some quantities are also produced in microbodies (Fig. 2). The scavenging of 0^{2-} by superoxide dismutase (SOD) results in the production of H_2O_2, which is removed by APX or CAT. However, both 0^{2-} and H_2O_2 are not as toxic as the (OH-), which is formed by the combination of O^2- and H_2O_2 in the presence of trace amounts of Fe^{2+} and Fe^{3+} by the Haber-Weiss reaction. The OH- can damage chlorophyll, protein, DNA, lipids and other important macromolecules, thus fatally affecting plant metabolism and limiting growth and yield (Sairam and Tyagi, 2004).

As depicted in Fig. 3, plants have developed a series of both enzymatic and non-enzymatic detoxification systems to counteract AOS, thereby protecting cells from oxidative damage (Sairam and Tyagi, 2004). Decrease in antioxidant activity in stressed tissues results in higher levels of AOS that may contribute to injury (Fadzillah *et. al.*, 1996). Protection against oxidative stress is an important component in determining the survival of a plant under temperature stress. Expression of stress proteins is an important adaptation to cope with environmental stresses. Most of the stress proteins are soluble in water and therefore contribute to stress tolerance presumably via hydration of cellular structures (Wahid and Close, 2007).

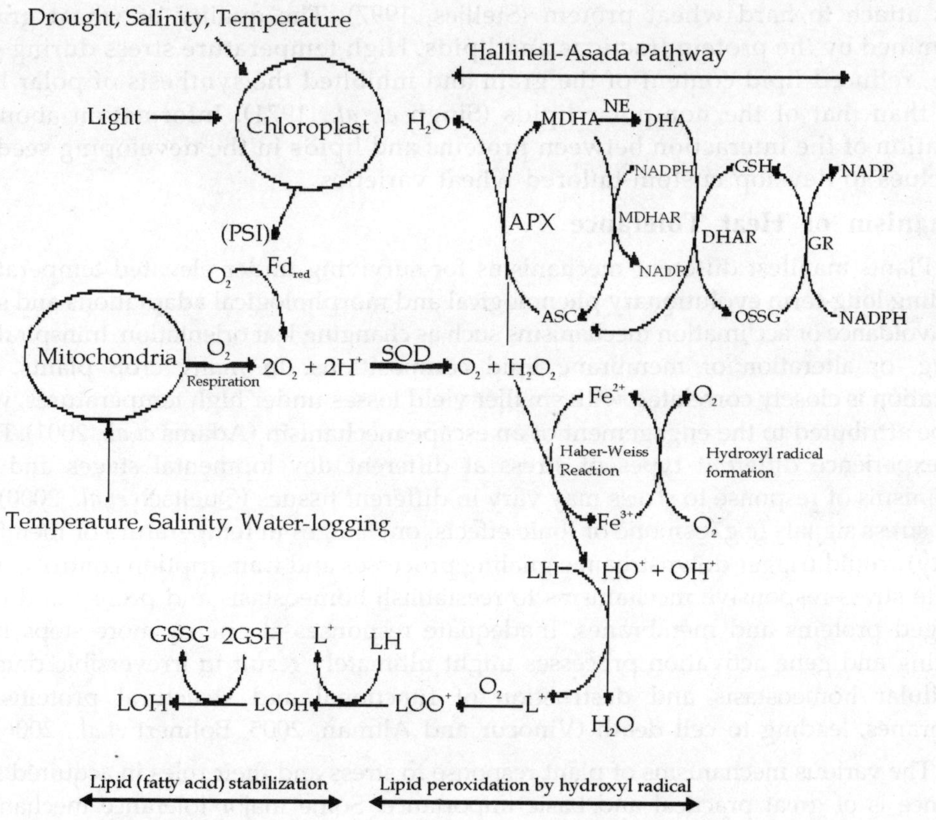

Fig. 3 : Schematic presentation for generation and scavenging of superoxide radical, hydrogen peroxide, hydroxyl radical-induced lipid peroxidation and glutathioneperoxidase-mediated fatty acid stabilization under environmental stresses. APX, ascorbate peroxidase; ASC, ascorbate; DHA, dehydroascorbate; DHAR, dehydroascorbatereductase; Fd, ferredoxin; GR, glutathione reductase; GSH, red glutathione; GSSG, oxi-glutathione; HO, hydroxyl radical; LH, lipid; L, LOO; LOOH,unstable lipid radicals and hydroperoxides; LOH, stable lipid (fatty acid); MDHA, monodehydro-ascorbate; MDHAR, mono dehydro-ascorbate reductase; NE,non-enzymatic reaction; PHGPX, phospholipid-hydroperoxide glutathione peroxidase; SOD, superoxide dismutase. Reproduced with permission from Sairam and Tyagi (2004).

Grain quality. Temperature stress not only affects growth and yield of the crops, but modifies the grain quality. Negative correlation between dough strength and hours above 35°C during grain filling in wheat suggested poor bread making property (Blumenthal *et. al.*, 1990). High temperature stress adversely affected UDP sucrose synthase enzyme in developing grains of barley, causing irreversible reduction in the capacity of the endosperm to convert sucrose to starch. Starch of non-waxy rice comprises of amylose and amylopectin, the ratio of which determines its cooking quality. Temperature stress affects the formation and development of the grain and synthesis of amylose and amylopectin (Umemoto *et. al.*, 1995). A group of proteins called friabilins binds to the polar lipids in the soft wheat, whereas none of these lipids attach to hard wheat protein (Stelljes, 1997). The quality of wheat grain is determined by the protein fractions and lipids. High temperature stress during grain filling, reduced lipid content of the grain and inhibited the synthesis of polar lipids more than that of the non-polar lipids (Singh *et. al.*, 1971). Information about the regulation of the interaction between proteins and lipids in the developing seed will give clues to develop custom-tailored wheat varieties.

Mechanism of Heat Tolerance

Plants manifest different mechanisms for surviving under elevated temperatures, including long-term evolutionary phenological and morphological adaptations and short-term avoidance or acclimation mechanisms such as changing leaf orientation, transpirational cooling, or alteration of membrane lipid compositions. In many crop plants, early maturation is closely correlated with smaller yield losses under high temperatures, which may be attributed to the engagement of an escape mechanism (Adams *et.al.*, 2001). Plants may experience different types of stress at different developmental stages and their mechanisms of response to stress may vary in different tissues (Queitsch *et.al.*, 2000). The initial stress signals (e.g. osmotic or ionic effects, or changes in temperature or membrane fluidity) would trigger downstream signaling processes and transcription controls, which activate stress-responsive mechanisms to reestablish homeostasis and protect and repair damaged proteins and membranes. Inadequate responses at one or more steps in the signaling and gene activation processes might ultimately result in irreversible damages in cellular homeostasis and destruction of functional and structural proteins and membranes, leading to cell death (Vinocur and Altman, 2005; Bohnert *et.al.*, 2006).

The various mechanisms of plant response to stress and their roles in acquired stress tolerance is of great practical and basic importance. Some major tolerance mechanisms, including ion transporters, osmoprotectants, free-radical scavengers, late embryogenesis abundant proteins and factors involved in signaling cascades and transcriptional control are essentially significant to counteract the stress effects (Wang *et. al.*, 2004). Series of changes and mechanisms, beginning with the perception of heat and signaling and production of metabolites that enable plants to cope with adversaries of heat stress, have been proposed (Fig. 4). Heat stress effects are notable at various levels, including plasma membrane and biochemical pathways operative in the cytosol (Sung *et.al.*, 2003). Initial effects of heat stress, however, are on plasmalemma, which shows more fluidity of lipid bilayer under stress. This leads to the induction of Ca^{2+} influx and cytoskeletal reorganization, resulting in the upregulation of mitogen acti-vated protein kinases and calcium dependent protein kinase.

Signaling of these cascades at nuclear level leads to the production of antioxidants and compatible osmolytes for cell water balance and osmotic adjustment. Production of ROS in the organelles (e.g., chloroplast and mitochondria) is of great significance for signaling as well as production of antioxidants (Bohnert *et.al.,* 2006). The antioxidant defense mechanism is a part of heat-stress adaptation, and its strength is correlated with acquisition of thermotolerance (Maestri *et.al.,* 2002). Accordingly, in a set of wheat genotypes, the capacity to acquire thermotolerance was correlated with activities of CAT and SOD, higher ascorbic acid content, and less oxidative damage (Sairam and Tyagi, 2004).

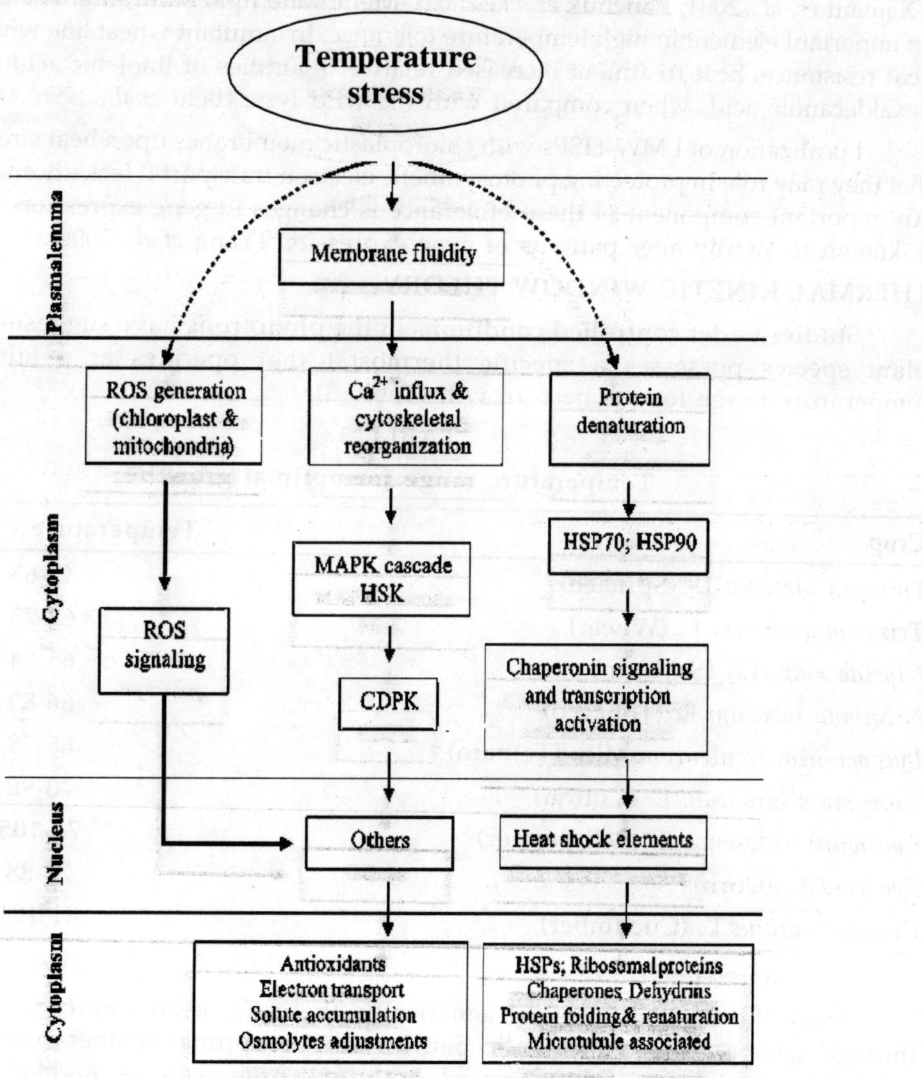

Fig. 4 : Proposed heat-stress tolerance mechanisms in plants. MAPK, mitogen activated protein kinases; ROS, reactive oxygen species; HAMK, heat shock activatedMAPK; HSE, heat shock element; HSPs, heat shock proteins; CDPK, calcium dependent protein kinase; HSK, histidine kinase. Partly adopted from Sung *et al.* (2003).

One of the most closely studied mechanisms of thermotolerance is the induction of HSPs, comprise several evolutionarily conserved protein families. However, each major HSP family has a unique mechanism of action with chaperonic activity. The protective effects of HSPs can be attributed to the network of the chaperone machinery, in which many chaperones act in concert. The HSPs/chaperones can play role in stress signal transduction and gene activation (Nollen and Morimoto, 2002). They also interact with other stress-response mechanisms such as production of osmolytes and antioxidants (Diamant *et. al.*, 2001; Panchuk *et. al.*, 2002). Membrane lipid saturation is considered as an important element in high temperature tolerance. In a mutant wheat line with increased heat resistance, heat treatment increased relative quantities of linolenic acid and *trans*3-hexaldecanoic acid, when compared with the wild type (Behl *et.al.*, 1996).

Localization of LMW-HSPs with chloroplastic membranes upon heat stress suggests that they play role in protecting photosynthetic electron transport (Heckathorn *et.al.*, 1998). An important component of thermotolerance is changes in gene expression. Heat stress is known to swiftly alter patterns of gene expression (Yang *et.al.*, 2006).

THERMAL KINETIC WINDOW THEORY

Studies under controlled conditions in the phytotrons have suggested that each plant species possesses a specific thermostat that operates at a fairly narrow temperature range for the best growth (Table 3).

TABLE 3

Temperature range for optimal growth

Crop	Temperature (°F) range
Spinacea oleracea L. (Spinach)	45-63
Triticum aestivum L. (Wheat)	62-73
Glycine max (L.) (Soybean)	65-84
Nicotiana tabacum L. (Tobacco)	66-82
Lycopersicon esculentum Mill. (Tomato)	68-78
Gossypium hirsutum L. (Cotton)	73-90
Capsicum frutesence L. (Bell pepper)	74-105
Zea may L. (Corn)	77-88
Cucumis sativus L. (Cucumber)	81-101

Source : Comis (1992)

Since the thermostat is set genetically, there lies a possibility for resetting it through gene-regulated metabolic pathways. The optimal temperature range of a particular species may shift toward both lower as well as higher side of the temperature regimes (Comis, 1992). Transfer of cucumber gene (optimal temperature 8] -101OF) to tobacco (optimal temperature 66-82°F) is linked with the production of the warm version of thermostable enzymes-hydroxypyruvate reductase and glutathione reductase for the protection of plants from heat stress. Since the altered tobacco plants also retained their cool version of the enzyme, they showed a combined temperature range (66°-101°F). The concept of thermal kinetic window theory, thus, proposes to produce tailor-made crop varieties with wider temperature range adaptation-under different agro-ecosystems. There lies a scope in testing this concept for improving productivity of crops under tropical and sub-tropical agro-climatic regions. Crops raised in the tropical areas (within a latitudinal range of 22.5° from the equator) represent the ideal system to study their responses to cool weather as well as warm weather.

Tropical refers to the regions which for a substantial part of the year, have warm weather (average temperature >17°C) and mostly extended frost free periods during which species tolerant of cool weather can be cropped. Amongst various biotic and abiotic constraints, temperature stresses are the most limiting factor for the productivity of the tropical grain legumes (Jayasundra *et .al.*, 1998). Warm season grain legumes are mainly cultivated in temperature Mediterranean and subtropical regions, but are also grown at high altitude in the tropics (Ethiopia). In comparison to grain legumes including many genera and species marked for warm season and cold season, respectively, wheat is being cultivated under wide climatic conditions, encompassing temperature, tropical and Mediterranean regions. Wheat is sensitive to low temperature for weed germination and seedling establishment (Singh and Gill 1975) and more sensitive to freezing damage as it enters into reproductive stage, particularly the period from ear emergence to two weeks -2°C in early spring, killing the earheads/ or restriciting photosynthesis and the movement of assimilates from the stem (Loss, 1989).

In the mediterranean environments, high temperature stress, complex with dryland conditions, is often an important constraint for wheat growth. Periods of high temperature near anthesis can dramatically reduce wheat yields.

Temperature above 30°C during reproductive differentiation caused floral sterility and reduced considerably grain tilling duration, resulting into shriveling of grains (Wardlaw *et.al.*, 1989, Saini and Lalonde 1998). Breeding wheat varieties for high temperature tolerance during reproductive differentiation and grain formation stage, is envisaged for increasing wheat yield in the mediterranean environment (Loss and Siddique 1994).

Genetics of Crop Thermal Kinetic

Genetic dissection of the morphological, biochemical and physiological parameters linked with temperature stress tolerance, can be performed by means by molecular markers, i.e. RFLP, RAPD, AFLP, microsatellites, etc. (Grillo and Leone 1997). These markers are genetically neutral in the sense that they are independent of gene expression of environmental effects or of interaction with other genes and show no temporal or tissue specificity. Their number is almost unlimited, so as to saturate the whole genome of a taxon. The availability of many species of densely saturated maps allows, atleast theoretically, the identification and chromosomal localization of nearly all the quantitative traits loci (QTL), controlling quantitative trait, and as thermotolerance components, on the basis of the association between the trait expression and the allelic composition of a segregating population at each of the marker loci included in the map (Tanksley, 1993).

The data so obtained can be subsequently pooled together for the evaluation of each genotype for global temperature stress tolerance and for the construction of the genomic architecture of the trait. Also, the information can be subsequently pooled together for the evaluation of each genotype for global temperature stress tolerance and for the construction of the genomic architecture of the trait. Also, the information can be exploited for marker assisted selection (MAS). The choice of the individuals to be included in the next generation is made in a non-destructive way on the basis of molecular alleles without the need to evaluate the character and at a very early stage of the plant development.

CONCLUSION

Plants exhibit a variety of responses to high temperatures, which are depicted by symptomatic and quantitative changes in growth and morphology. The ability of the plant to cope with or adjust to the temperature stress varies across and within species as well as at different developmental stages. Although high temperatures affect plant growth at all developmental stages, later phenological stages, in particular anthesis and grain filling, are generally more susceptible. Pollen viability, patterns of assimilate partitioning, and growth and development of seed/grain are highly adversely affected. Other notable temperature stress effects include structural changes in tissues and cell organelles, disorganization of cell membranes, disturbance of leaf water relations, and impedance of photosynthesis via effects on photochemical and biochemical reactions and photosynthetic membranes. Lipid peroxidation via the production of ROS and changes in antioxidant enzymes and altered pattern of synthesis of primary and secondary metabolites are also of considerable importance.

In response to temperature stress, plants manifest numerous adaptive changes. The induction of signaling cascades leading to profound changes in specific gene expression is considered an important heat-stress adaptation. Although various signaling molecules are synthesized under heat stress, the role of Ca^{2+} remains critical. A fundamental temperature-stress response ubiquitous to plants is the expression of HSPs, which range from low (10 kDa) to high (100 kDa) molecular mass in different species.Temperature is one of the most important abiotic factors, controlling the adaptation, acclimation, production and geographical distribution of the natural as well as the cultivated plant species. Plants have evolved various morphological, anatomical, biochemical and physiological means to cope with the undesirable temperature regimes ranging from freezing and below freezing point in the temperate regions to high temperature of the arid conditions in the desert. Crop cultivation is limited to temperature regimes operating in a narrow range of 10°-30°. Strategies are to be evolved for breeding low temperature and high temperature tolerant crops either by hybridization or genetic engineering techniques. In any situation, thorough understanding of the molecular, biochemical and physiological responses of the plants is a must. Regulation of temperature tolerance in plants seems primarily to be a function of membrane thermal stability and thermal protective devices such as cold tolerant proteins and heat shock proteins, which are monitored through temperature induced expression and function of the responsive genes. Temperature tolerance is a polygenic character and many genes have been characterized, imparting respective low temperature and high temperature tolerance in transgenic plants, especially *Arabidopsis thaliana*. Recent studies on common switch (control factor) regulating the expression of a set of multiple temperature responsive genes, open up the new thinking confirming about the unified theory of temperature tolerance in plants in general and crop plants in particular. Enhancing the temperature tolerance capacity across the low as well as higher temperature limits, can be exploited for extending the cultivation beyond the traditional areas where present crop varieties are being grown.

The physiological mechanisms of heat tolerance are relatively well understood, further studies are essential to determine physiological basis of assimilate partitioning from source to sink, plant phenotypic flexibility which leads to heat tolerance, and factors that modulate plant heat-stress response. Furthermore, applications of genomics, proteomics and trascriptomics approaches to a better understanding of the molecular basis of plant response to heat stress as well as plant heat tolerance are imperative. As in the case of most other abiotic stresses, foliar plant parts are more directly impinged upon by high temperatures than roots. However, an understanding of root responses to heat stress, most likely involving root-shoot signaling, is crucial and warrants further exploration. Molecular knowledge of response and tolerance mechanisms will pave the way for engineering plants that can tolerate temperature stress and could be the basis for production of crops which can produce economic yield under temperature-stress conditions.

REFERENCES

Adams, S.R., Cockshull, K.E., Cave, C.R.J., 2001. Effect of temperature on the growth and development of tomato fruits. *Ann. Bot.* 88, 869-877.

Ahn, Y.-J., Zimmerman, J.L., 2006. Introduction of the calTot HSP17.7 into potato *(Solanum tuberosum* L.) enhances cellular membrane stability and tuberization in vitro. *Plant Cell Environ.* 29, 95-104.

Anon, S., Fernandez, J.A., Franco, J.A., TOlTecillas, A., Alarcon, J.1., Sanchez-Blanco, M.J., 2004. Effects of water stress and night temperature preconditioning on water relations and morphological and anatomical changes of *Lotus creticus* plants. *Sci. Hortic.* 101, 333-342.

Antikainen, M. and Grifith, M. 1997. Antifreeze protein accumulation in freezing-tolerant cereals. *Physiol. Plant* **97** : 423-432.

Antunes, M.D.C., Sfakiotakis, E.M., 2000. Effect of high temperature stress on ethylene biosynthesis, respiration and ripening of 'Hayward' kiwifruit. *Postharvest BioI. Technol.* 20, 251-259.

Arora, R., Rawland, L. and Panta, G.R. 1997. Chill-responsive dehydrins in blueberry: Are they associated with cold hardiness or dormancy transition? *Physiol. Plant* **101** : 8-16.

Arshad, M., Frankenberger, W.T.1., 2002. Ethylene, Agricultural Sources and Applications. Kluwer Academic/Plenum Publishers, New York.

Ashraf, M., Foolad, M.R., 2007. Roles of glycine betaine and proline in improving plant abiotic stress resistance. *Environ. Exp. Bot.* 59, 206-216.

Ashraf, M., Hafeez, M., 2004. Thermotolerance of pearl millet and maize at early growth stages: growth and nutrient relations. *Biol. Plant.* 48, 81-86.

Bannu EFEOGLU (2009). Heat shock proteins and heat shock response in plants. *G.U. Journal of science.* 22(2):67-75.

Banowetz, G.M., Ammar, K., Chen, D.D., 1999. Temperature effects on cytokinin accumulation and kernel mass in dwmf wheat. *Ann. Bot.* 83, 303-307.

Barua, D., Downs, c.A., Heckathorn, S.A., 2003. Variation in chloroplast small heat-shock protein function is a major determinant of variation in thermotolerance of photosynthetic electron transport among ecotypes of *Chenopodium album. Funct. Plant Biol.* 30, 1071-1079.

Behl, R.K., Heise, K.P., Moawad, A.M., 1996. High temperature tolerance in relation to changes in lipids in mutant wheat. Der Tropenlandwirt 97, 131-135.

Berry, J.A. and Raison, J.K. 1981. Response of macrophytes to temperature. In : Q.L. Lange, P.S. Nobel, C.B. Osmond and H. Zeigler (eds). *Encyclopedia of Plant Physiology,* Vol. 12A. Springer-Verlag, Berlin, pp 277-388.

Bewley, J.D. 1979. Physiological aspects of desiccation tolerance. *A. Rev. Pl. Physiol.* **30** : 195-230.

Bianchi, G., Gamba, A., Limiroli, R., Pozzi, N., Elster, R., Salamini, F. and Bartel, D. 1993. The unusual sugar composition in the leaves of the resurrection plant, *Myrothamnus flabellifolia. Physiol. Plant* **87** : 223-226.

Bienz, M. and Pellham, H.R.B. 1987. Mechanism of heat-shock gene activation in higher eukaryotes. *Adv. Genetics* **24** : 31-72.

Blumenthal, C., Barlow, S. and Wrigley, S. 1990. Global warming and wheat. *Nature* **347** : 325.

Bohnert, H.1., Gong, Q., Li, P., Ma, S., 2006. Unraveling abiotic stress tolerance mechanisms-getting genomics going. CUIT. Opin. Plant BioI. 9,180-188.

Bowen, J., Michael, L.-Y., Plummer, K.1.M., Ferguson, 1.A.N., 2002. The heat shock response is involved in thermotolerance in suspension-cultured apple fruit cells. J. Plant Physiol. 159, 599-606.

Bowers, M.C. 1994. Environmental effects of cold on plants. In : R.E. Wilkinson (eds). *Plant Environment Interactions*. Marcel Dekkar, New York, pp. 31-34.

Brock, T.D. 1985. Life at high temperatures. *Science* 230 : 132-138.

Camejo, D., Rodriguez, P., Morales, M.A., Dell' amico, J.M., TOlTecillas, A., Alarcon, J.J., 2005. High temperature effects on photosynthetic activity of two tomato cultivars with different heat susceptibility. *J. Plant Physiol.* 162, 281-289.

Coleman, J.S. and McConnaughay, K. 1994. Phenotypic plasticity. *Trends Ecol. Evol.* **9** : 187-191.

Comis, D. 1992. Resetting a plant's thermostat. *USDA Agric. Res.* **40** : 14-16.

De Ronde, J.AD., Cress, W.A, Kruger, G.H.J., Strasser, RJ., Staden, J.V., 2004. Photosynthetic response of transgenic soybean plants containing an *Arabidopsis P5CR* gene, during heat and drought stress. *J. Plant Physiol.* 61, 1211-1244.

Dhaliwal, G.S. and Kler, D.S. 1995. *Principles of Agricultural Ecology*. Himalaya Publishing House, Bombay.

Dhaubhadel, S., Chaudhary, S., Dobinson, K.F., Krishna, P., 1999. Treatment with 24-epibrassinolide, a brassinosteroid, increases the basic thermotolerance of *Brassica napus* and tomato seedlings. Plant Mol. Biol. **40**, 333–342.

Diamant, S., Eliahu, N., Rosenthal, D., Goloubinoff, P., 2001. Chemical chaperones regulate molecular chaperones in vitro and in cells under combined salt and heat stresses. *J. BioI. Chern.* 276, 39586-39591.

Fadzillah, N.M., Gill, V., Finch, R.P., Burdon, R.H., 1996. Chilling, oxidative stress and antioxidant responses in shoot cultures of rice. *Planta* **199**, 552–556.

Fenglu, Z., Ming, Z., Zhimin, W., Shuan, W.ZJ., Jinglum, G., 1997. Maize kernel development and release of ethylene by grains. J. China Agric. Univ. 2,85-89.

Feussner, K., Feussner, 1., Leopold, 1., Wasternack, C., 1997. Isolation of a cDNA coding for an ubiquitin-conjugating enzyme UBC1 of tomato-the first stress-induced UBC of higher plants. FEBS Lett. 409, 211215.

Foolad, M.R., 2005. Breeding for abiotic stress tolerances in tomato. In: Ashraf, M., Harris, PJ.C. (Eds.), Abiotic Stresses: Plant Resistance Through Breeding and Molecular Approaches. The Haworth Press Inc., New York, USA, pp.613-684.

Gaff, D.F. 1986. Desiccation tolerant resurrection grasses from Kenya and West Africa. *Oecologia* **70** : 118-120.

Giaveno, C., FelTero, J., 2003. Introduction of tropical maize genotypes to increase silage production in the central area of Santa Fe, Argentina. Crop Breed. Appl. Biotechnol. 3, 89-94.

Graham, D. and Patterson, B.D. 1982. Response of plant to low, non-freezing temperature, proteins, metabolism and acclimation. *A. Rev. Pl. Physiol.* **33** : 347-372.

Grillo, S. and Leone, A. 1997. *Physical Stress in Plants: Genes and Their Products for Their Tolerance*. Springer Verlag, Berlin.

Guy, C.L. 1990. Cold acclimation and freezing stress tolerance: Role of protein metabolism. *A. Rev. Pl. Physiol. Pl. Mol. Biol.* **41** : 187-223.

Hajela, R.K., Horvath, D.P., Gilmour, S.J. and Thomashaw, M.F. 1990. Molecular cloning and expression of Cor (cold-regulated) genes in *Arabidopsis thaliana*. *Pl. Physiol.* **93** : 1246-1252.

Hall, AE., 2001. Crop Responses to Environment. CRC Press LLC, Boca Raton, Florida.

Heckathorn, S.A, Downs, C.A, Coleman, J.S., 1998. Nuclear-encoded chloroplast proteins accunulated in the cytosol during severe heat stress. Intern. J. Plant Sci. 159,39-45.

Holappa, L.D. and Walker-Simmons, M.K. 1995. The wheat abscisic acid-responsiveness protein kinase mRNA, PKAB is up regulated by dehydration, cold temperature and osmotic stress. *Pl. Physiol.* **108** : 1203-1210.

Howarth, CJ., 2005. Genetic improvements of tolerance to high temperature. In: Ashraf, M., HalTis, PJ.C. (Eds.), Abiotic Stresses: Plant Resistance Through Breeding and Molecular Approaches. Howarth Press Inc., New York.

Hughes, M.A. and Dunn, M.A. 1996. The molecular biology of plant acclimation to low temperature. *J. Expt. Bot.* **47** : 291-305.

Iba, K., 2002. Acclimative response to temperature stress in higher plants: approaches of gene engineering for temperature tolerance. Annu. Rev. Plant Biol. **53**, 225–245.

Ingram, J. and Bartel, D. 1996. The molecular basis of dehydration tolerance in plants. *A. Rev. Pl. Physiol. Pl. Mol. Biol.* **47** : 377.

Jaglo-Ottosen, K.R., Gilmour, S.J., Zarka, D.G., Schabenberger, O. and Thomashow, M.F. 1998. *Arabidopsis* CBF1 overexpression induces Cor gene and enhances freezing tolerance. *Science* **280** : 104-106.

Jayasundara, H.P.S., Thompson, B.D. and Tang, C. 1998. Response of cool season grain legumes to soil abiotic stress. *Adv. Agron.* **63** : 77-151.

Jiang, C., Lu, B. and Singh, B. 1996. Requirement of CCGAC *Cis*-acting element for cold induction to soil abiotic stresses. *Adv. Agron.* **63** : 77-151.

Kavi Kishore, P.B., Sangam, S., Amrutha, R.N., Laxmi, P.S., Naidu, K.R., Rao, K.R.S.S., Rao, S., Reddy, KJ., Theriappan, P., Sreenivasulu, N., 2005. Regulation of proline biosynthesis, degradation, uptake and transport in higher plants: its implications in plant growth and abiotic stress tolerance. CUlT. Sci. 88, 424-438.

Kawano, T., Sahashi, N., Takahashi, K., Uozumi, N., Muto, S., 1998. Salicylic acid induces extracellular superoxide generation followed by an increase in cytosolic calcium ion in tobacco suspension culture: the earliest events in salicylic acid signal transduction. Plant Cell Physiol. 39, 721730.

Kepova, K.D., Holzer, R., Stoilova, L.S., Feller, U., 2005. Heat stress effects on ribulose-1,5-bisphosphate carboxylase/oxygenase, Rubisco bindind protein and Rubisco activase in wheat leaves. Biol. Plant. 49, 521-525.

Kim, K.K., Kim, R. Kim, S.H. 1998. Crystal structure of a small heat shock protein. *Nature* **394** : 595-599.

Kodama, H., Horiguchi, G., Nishiushi, T., Nishimura, M. and Iba, K. 1995. Fatty acid desaturation during chilling acclimation is one of the factors involved in conferring low-temperature tolerance to young tobacco leaves. *Pl. Physiol.* **107** : 1177-1185.

Kojma, M., Suzuki, H., Ohmishi, M. and Ito, S. 1998. Effect of growth temperature on lipids of Adzuki bean cells. *Phytochemistry* **47** : 1483-1487.

Korotaeva, N.E., Antipina, A.1., Grabelynch, 0.1., Varakina, N.N., Borovskii, G.B., Voinikov, V.K., 2001. Mitochondrial low-molecular-weight heat shock proteins and tolerance of crop plant's mitochondria to hyperthermia. Fiziol. Biokhim Kul'turn. Rasten. 29, 271-276.

Laloi, M., Klein, M., Riesmeier, J.W., Muller-Robber, B., Fleury, C., Bouillard, F. and Ricquier, D. 1997. A plant cold-induced uncoupling protein. *Nature* **389** : 135-136.

Lie, J.H., Hubel, A. and Schoffl, F. 1995. Depression of the activity of genetically engineered heat shock factor causes constitutive synthesis of heat shock proteins and increases thermotolerance in transgenic *Arabidopsis*. *Pl. J.* **8** : 603-612.

Lindquist, S. 1986. The heat shock response. *A. Rev. Biochem.* **55** : 1151-1191.

Liu, N., Ko, S., Yeh, K.-C., Charng, Y., 2006. Isolation and characterization of tomato Hsa32 encoding a novel heat-shock protein. Plant Sci. 170,976-985.

Liu, X., Huang, B., 2000. Heat stress injury in relation to membrane lipid peroxidation in creeping bent grass. Crop Sci. 40, 503-510.

Llusia, J., Penuelas, J., Munne-Bosch, S., 2005. Sustained accumulation of methyl salicylate alters antioxidant protection and reduces tolerance of holm oak to heat stress. Physiol. Plant. 124,353-361.

Loss, S.P. 1989. Frost injury to wheat. *J. Agric. West Aust.* **3** : 32-34.

Loss, S.P. and Siddique, K.H.M. 1994. Morphological and physiological traits associated with wheat yield increases in Mediterranean environment. *Adv. Agron.* **52** : 269-276.

Machado, S., Paulsen, G.M., 2001. Combined effects of drought and high temperature on water relations of wheat and sorghum. *Plant Soil,* 233.

Maestri, E., Klueva, N., PelTotta, C., Gulli, M., Nguyen, H.T., Marmiroli, N., 2002. Molecular genetics of heat tolerance and heat shock proteins in cereals. *Plant Mol. BioI.* 48, 667-681.

Mahajan, V., Dhillon, B.S., Khera, A.S. and Singh, O. 1993. Combining ability analysis response to cold stress in maize. *Field Crops Res.* **34** : 71-82.

Marchand, EL., Mertens, S., Kockelbergh, E, Beyens, L., Nijs, 1., 2005. Performance of high arctic tundra plants improved during but deteriorated after exposure to a simulated extreme temperature event. *Global Change BioI.* 11, 2078-2089.

Maresca, B. and Cossins, A.R. 1993. Fatty freedback and fluidity. *Nature* **365** : 606-607.

Mariama, M., Muthukumar, B., Veluthambi, K. and Gnanam, A. 1997. Effect of high temperature stress on the expression of low molecular weight heat shock proteins in rice leaves. *J. Pl. Physiol.* **151** : 763-765.

Matsuura-Enlo, C., Maeshima, M. and Yoshida, S. 1992. Mechanism of the decline in vascular H^+-ATPase activity in mungbean hypocotyls during chilling. *Pl. Physiol.* **100** : 718-722.

Mazorra, L.M., Nunez, M., Echerarria, E., Coll, F., S´anchez-Blanco, M.J., 2002. Influence of brassinosteriods and antioxidant enzymes activity in tomato under different temperatures. Plant Biol. **45**, 593–596.

McAinsh, M.R., Brownlee, C. and Hetherington, A.M. 1990. Abs*cisic* acid induced elevation of guard cell cytosolic Ca^{2+} precedes stomatal closure. *Nature* **343** : 186-188.

McKersie, B.D. and Ya'acov, Y.L. 1994. *Stress and Stress Coping in Cultivated Plants*. Kluwer Academic Publishers, Dordrecht, The Netherlands.

Miller, P., Lanier, W., Brandt, S., 2001. Using Growing Degree Days to Predict Plant Stages. AglExtension Communications Coordinator, Communications Services, Montana State University-Bozeman, Bozeman, MO. Moderate increase of mean daily temperature adversely affects fruit set of *Lycopersicon esculentum* by disrupting specific physiological processes in male reproductive development. *Ann. Bot.* 97, 731-738.

Maffat, J.M., Sears, R.G. and Paulson, G.M. 1990. Wheat high temperature tolerance during reproductive growth. I. Evaluation by chlorophyll fluorescence. *Crop Sci.* **30** : 881-885.

Morrison, M.J., Stewart, D.W., 2002. Heat stress during flowering in summer brassica. *Crop Sci.* 42, 797-803.

Momcilovic, 1., Ristic, Z., 2007. Expression of chloroplast protein synthesis elongation factor, EF-Tu, in two lines of maize with contrasting tolerance to heat stress during early stages of plant development. *J. Plant Physiol.* 164, 90-99.

Moore, P.D. 1998. Desert ecology: Life in the upper crest. *Nature* **393**: 419-420.

Morales, D., Rodriguez, P., Dell'amico, J., Nicolas, E., TOlTecillas, A., SanchezBlanco, M.J., 2003. High-temperature preconditioning and thermal shock imposition affects water relations, gas exchange and root hydraulic conductivity in tomato. *BioI. Plant.* 47, 203-208.

Morita, S., Siratsuchi, H., Takanashi, J., Fujita, K., 2004. Effect of high temperature on ripening in rice plant. Analysis of the effect of high night and high day temperature applied to the panicle in other parts of the plant. *Jpn. J. Crop. Sci.* 73, 77-83.

Munne-Bosch, S., Lopez-Carbonell, M., Alegre, L.A., Van Onckelen, H.A., 2002. Effect of drought and high solar radiation on 1-aminocyclopropane-1carboxylic acid and abscisic acid in *Rosmarinus officinalis* plants. *Physiol.* Plant. 114, 380-386.

Murai, M. and Yoshida, S. 1998. Vascuolar-membrane lesions induced by a freeze-thaw cycle in protoplasts isolated from deaclimated tubers of Jerusalem Artiochoke (*Helianthus tuberosus* L.). *Pl. Cell Physiol.* **39** : 87-96.

Murata, N., Ishizaki, N.O., Higashi, S., Hagashi, H., Taseka, Y. and Nishida, I. 1992. Genetically engineered alteration in chilling sensitivity in plants. *Nature* **356** : 710-713.

Nakamoto, H., Hiyama, T., 1999. Heat-shock proteins and temperature stress. In: Pessarakli, M. (Ed.), Handbook of Plant and Crop Stress. Marcel Dekker, New York, pp. 399-416.

Nascimento, W.M., Cantliffe, DJ., Huber, DJ., 2004. Ethylene evolution and endo-l3-mannanase activity during lettuce seed germination at high temperature. Scientia Agricola 61, 156-163.

NDong, C., Quellet, F., Houde, M. and Seshan, F. 1997. Gene expression during cold acclimation in strawberry. *Pl. Cell Physiol.* **38** : 863-870.

Neta-Sharir, 1., Isaacson, T., Lurie, S., Weiss, D., 2005. Dual role for tomato heat shock protein 21: protecting photosystem ii from oxidative stress and promoting color changes during fruit maturation. Plant Cell 17, 1829-1838.

Neumann, D.M., Emmermann, M., Thielfelder, J.M., Zur Nieden, U., Clericus, M., Braun, H.P., Nover, L., Schmitz, U.K., 1993. HSP68-a DNAK-like heat-stress protein of plant mitochondria. Planta 190, 32-43.

Nilsen, E.T. and Orcutt, D.M. 1996. Physiology of Plant Under Stress: Abiotic Factors. John Wiley and Sons, Inc., New York.

Nishida, I. and Murata, N. 1996. Chilling sensitivity in plants and cynobacteria. The crucial contribution of membrane lipids. *A. Rev. Pl. Physiol. Pl. Mol. Biol.* **47** : 541-568.

Nollen, E.A.A., Morimoto, R.1., 2002. Chaperoning signaling pathways: molecular chaperones as stress-sensing 'heat shock' proteins. *J. Cell Sci.* 115, 2809-2816.

Osteryoung, K.W. and Vierling, E. 1994. Conserved cell and organelle division. *Nature* **376** : 473-474.

Panchuk, 1.1., Volkov, R.A, Sch6ffl, E, 2002. Heat stress- and heat shock transcription factor-dependent expression and activity of ascorbate peroxidase in *Arabidopsis.* Plant Physiol. 129, 838-853.

Patel, P.N., Hall, AE., 1990. Genotypic variation and classification of cowpea for reproductive responses to high temperatures under long photoperiods. Crop Sci. 30, 614-621.

Pearce, R.S. and Ashworth, E.N. 1992. Cell shape and localization of ice in leaves of overwintering wheat during frost stress in the field. *Planta* **188** : 327-331.

Pearcy, R.W., Berry, J.A. and Fork, D.C. 1977. Effect of growth temperature on the thermal stability of photosynthetic apparatus of *Atriplex lentiformis* (Torr.) Wats. *Pl. Physiol.* **59** : 873-878.

Peet, M.M., Willits, D.H., 1998. The effect of night temperature on greenhouse grown tomato yields in warm climate. Agric. *Forest Meteorol.* 92, 191-202.

Pei, Z.M., Ghassemian, M., Kwak, C.M., McCourt, P. and Schroeder, J.I. 1998. Role of farnesyltransferase in ABA regulation of guard cell anion channels and plant water loss. *Science* **282** : 287-290.

Porter, J.R., 2005. Rising temperatures are likely to reduce crop yields. Nature 436,174.

Prasinos, C., Krampis, K., Samakovli, D., Hatzopoulos, P., 2005. Tight regulation of expression of two Arabidopsis cytosolic *Hsp90* genes during embryo development. J. Exp. Bot. 56, 633-644.

Psenner, R. and Sattler, B. 1998. Life at freezing point. *Science* **280** : 2073-2074.

Quan, R., Shang, M., Zhang, H., Zhao, Y., Zhang, J., 2004. Engineering of enhanced glycine betaine synthesis improves drought tolerance in maize. Plant Biotech. J. 2, 477-486.

Queitsch, C., Hong, S.W., Vierling, E., Lindquest, S., 2000. Heat shock protein 101 plays a crucial role in thermotolerance in *Arabidopsis*. Plant Cell 12, 479-492.

Rainey, K., Griffiths, P., 2005. Evaluation of *Phaseolus acutifolius* A Gray plant introductions under high temperatures in a controlled environment. Genet. Resour. Crop Evol. 52, 117-120.

Rehman, H., Malik, S.A, Saleem, M., 2004. Heat tolerance of upland cotton during the fruiting stage evaluated during cellular membrane thermostability. Field Crops Res. 85, 149-158.

Rekika, D., Monneveux, P. and Havaux, M. 1997. In vivo tolerance of photosynthetic membrane to high and low temperature in cultivated and wild wheat of *Triticum* and *Aegilops* genes. *J. Pl. Physiol.* **150** : 734-738.

Saini, H.S. and Lalonde, S. 1998. Injuries to reproductive development under water stress and their consequences for crop productivity. *J. Crop Prod.* **1** : 223-248.

Sairam, R.K., Tyagi, A, 2004. Physiology and molecular biology of salinity stress tolerance in plants. CUIT. Sci. 86, 407-421.

Sakamoto, A, Murata, N., 2002. The role of glycine betaine in the protection of plants from stress: clues from transgenic plants. Plant Cell Environ. 25, 163-171.

Salahas, G., Peralis, B., Georgion, C.D. and Cavalas, N.A. 1997. Trehalose, an extreme temperature protector of phosphoenolpyruvate carboxylase from the C_4-plant Cynadon dactylon. *Phytochemistry* **46** : 133-134.

Salvucci, M.E., Crafts-Brandner, S.1., 2004b. Inhibition of photosynthesis by heat stress: the activation state of Rubisco as a limiting factor in photosynthesis. Physiol. Plant. 120, 179-186.

Sato, S., Kamiyama, M., Iwata, T., Makita, N., Furukawa, H., Ikeda, H., 2006.

Schffl, E, Prandl, R., Reindl, A, 1999. Molecular responses to heat stress. In: Shinozaki, K., Yamaguchi-Shinozaki, K. (Eds.), Molecular Responses to Cold, Drought, Heat and Salt Stress in Higher Plants. R.G. Landes Co., Austin, Texas, pp. 81-98.

Seemann, J.R., Downton, W.J.S. and Berry, J.A. 1986. Temperature and leaf osmotic potential as factor in the acclimation of photosynthesis to high temperature in desert plants. *Pl. Physiol.* **80** : 926-930.

Sharkova, V.E., 2001. The effect of heat shock on the capacity of wheat plants to restore their photosynthetic electron transport after photoinhibition or repeated heating. Russ. J. Plant Physiol. 48, 793-797.

Sheen, J. 1996. Ca^{2+}-dependent protein kinases and stress signal transduction in plants. *Science* **274** : 1900-1902.

Siddique, K.H.M., Loss, S.P., Regan, K.L., Jettner, R.L., 1999. Adaptation and seed yield of cool season grain legumes in MeditelTanean environments of south-western Australia. Aust. J. Agric. Res. 50, 375-387.

Simoes-Araujo, J.L., Rumjanek, N.G., Margis-Pinheiro, M., 2003. Small heat shock proteins genes are differentially expressed in distinct varieties of common bean. Braz. J. Plant Physiol. 15, 33-41.

Singh, O.S. and Gill, K.S. 1975. Some physiological aspects of effects of kinetin, depth of seeding and soil temperature on seedling establishment and metabolism of dwarf wheat. *Indian J. Agric. Sci.* **42** : 209-215.

Singh, O.S., Gill, D.S. and Dhaliwal, G.S. 1997. Impact of air pollution on agro-ecosystem. *Indian J. Ecol.* **24** : 136-156.

Singh, O.S., Sukhija, P.S., Nagpal, M.L. and Bhatia, I.S. 1971. Effect of soil moisture stress on the lipid quality of matured wheat grains. *Indian J. agric. Sci.* **41** : 300-304.

Singla, S.L., Pareak, A. and Grover, A. 1996. High temperature. In: M.N.V. Prasad (eds.) *Plant Ecophysiology*. John Wiley and Sons, Inc. New York, pp. 101-127.

Sakamoto, A., Murata, N., 2002. The role of glycine betaine in the protection of plants from stress: clues from transgenic plants. Plant Cell Environ. **25**, 163–171.

Stelljes, K.B. 1997. New clues to wheat hardiness. *Agric. Res.* (USDA) **45** : 21.

Stone, P.1., Nicolas, M.E., 1994. Wheat cultivars vary widely in their responses of grain yield and quality to short periods of post-anthesis heat stress. Aust. J. Plant Physiol. 21, 887-900.

Sung, D.-Y., Kaplan, E, Lee, K.-J., Guy, C.L., 2003. Acquired tolerance to temperature extremes. Trends Plant Sci. 8, 179-187.

Taiz, L., Zeiger, E., 2006. Plant Physiology. Sinauer Associates Inc. Publishers, Massachusetts.

Tanksley, S.D. 1993. Mapping polygenes. *A. Rev. Genet.* **27**: 205-233.

Tewolde, H., Fernandez, C.J., Erickson, C.A, 2006. Wheat cultivars adapted to post-heading high temperature stress. J. Agron. Crop Sci. 192, 111-120.

Todorov, D.T., Karanov, E.N., Smith, A.R., Hall, M.A., 2003. Chlorophyllase activity and chlorophyll content in wild type and *eti 5* mutant of *Arabidopsis thaliana* subjected to low and high temperatures. Biol. Plant. **46**, 633–636.

Thompson, L.M., 1986. Climatic change, weather variability and com production. Agron. J. 78, 649-653.

Toth, SZ., Schansker, G., Kissimon, J., Kovacs, L., Garab, G., Strasser, RJ., 2005. Biophysical studies of photosystem II-related recovery processes after a heat pulse in barley seedlings *(Hordeum vulgare* L.). J. Plant Physiol. 162, 181-194.

Tsukaguchi, T., Kawamitsu, Y., Takeda, H., Suzuki, K., Egawa, Y., 2003. Water status of flower buds and leaves as affected by high temperature in heat tolerant and heat-sensitive cultivars of snap bean *(Phaseolus vulgaris* L.). Plant Prod. Sci. 6, 4-27.

Uemura, M., Joseph, R.A. and Steponkus, P.L. 1995. Cold acclimation of *Arabidopsis thaliana*. Effect on plasma membrane lipid composition and freeze-induced lesions. *Pl. Physiol.* **109** : 15-13.

Umemoto, T., Nakamura, Y. and Ishikura, N. 1995. Activity of starch synthesis and amylase content in rice endosperm. *Phytochemistry* **40** : 1613-1616.

Vara Prasad, P.V., Craufurd, P.Q., Summerfield, RJ., 1999. Fruit number in relation to pollen production and viability in groundnut exposed to short episodes of heat stress. Ann. Bot. 84, 381-386.

Vettakkorumakankav, N.N., Falk, D., Saxena, P., Fletcher, R.A, 1999. A crucial role for gibberellins in stress protection of plants. Plant Cell Physiol. 40, 542-548.

Veziana, L.P., Ferullo, J.M., Laliberte, G., Laberge, S. and Willemot, C. 1996. Chilling and freezing. In : M.N.V. Prasad (ed.). *Plant Ecophysiology*. John Wiley and Sons, Inc., New York, pp. 61-100.

Vigh, L., Gombos, Z., Horvath, I. and Joo, F. 1993. Saturation of membrane lipids by hydrogenation induces thermal stability in chloroplast inhibiting the heat dependent stimulation of photosystem. I. Mediated electron transport. *Biochem. Biophy. Acta* **979** : 361-364.

Vinocur, B., Altman, A, 2005. Recent advances in engineering plant tolerance to abiotic stress: achievements and limitations. CUIT. Opin. Biotechnol. 16, 123-132.

Vollenweider, P., Gunthardt-Goerg, M.S., 2005. Diagnosis of abiotic and biotic stress factors using the visible symptoms in foliage. Environ. Pollut. 137, 455-465.

Wahid, A, 2007. Physiological implications of metabolites biosynthesis in net assimilation and heat stress tolerance of sugarcane sprouts. J. Plant Res. 120, 219-228.

Wahid, A., Ghazanfar, A., 2006. Possible involvement of some secondary metabolites in salt tolerance of sugarcane. J. Plant Physiol. **163**, 723–730

Wahid, A, Close, TJ., 2007. Expression of dehydrins under heat stress and their relationship with water relations of sugarcane leaves. Bio. Plant. 51, 104-109.

Wang, D., Luthe, D.S., 2003. Heat sensitivity in a bentgrass variant: failure to accumulate a chloroplast heat shock protein isoform implicated in heat tolerance. Plant Physiol. 133, 319-327.

Wang, H., Dalta, R., Georges, F., Lowen, M. and Cultler, A.J. 1995. Promoters from Kin 1 and Cor 6.6, two homologus *Arabidopsis thaliana* genes: transcriptional regulation and gene expression induced by low temperature, ABA, osmoticum and dehydration. Pl. Mol. Biol. **28** : 605-617.

Wang, L.-J., Li, S.H., 2006a. Thermotolerance and related antioxidant enzyme activities induced by heat acclimation and salicylic acid in grape (*Vitis vinifera* L.) leaves. Plant Growth Regul. 48, 137-144.

Wang, L.-J., Li, S.-L., 2006b. Salicylic acid-induced heat or cold tolerance in relation to Ca^2+ homeostasis and antioxidant systems in young grape plants. Plant Sci. 170, 685-694.

Wang, W., Vinocur, B., Shoseyov, 0., Altman, A, 2004. Role of plant heat-shock proteins and molecular chaperones in the abiotic stress response. Trends Plant Sci. 9, 244-252.

Wardlaw, 1.E, 1974. Temperature control of translocation. In: Bielske, R.L., Ferguson, AR., Cresswell, M.M. (Eds.), Mechanism of Regulation of Plant Growth. Bull. Royal Soc., New Zealand, Wellington, pp. 533-538.

Wardlaw, I.F., Dawson, I.A. and Munibi, P. 1989. The tolerance of wheat to high temperature during reproductive growth. II. Grain development. Aust. J. agric. Res. **40** : 15-24.

Weis, E. and Berry, J.A. 1988. Plant and high temperature stress. In : S.P. Long and F.I. Woodward (eds). Plant and Temperature. The Company of Biologists Ltd., Cambridge, UK, pp. 329-346.

Wise, R.R., Olson, AJ., Schrader, S.M., Sharkey, T.D., 2004. Electron transport is the functional limitation of photosynthesis in field-grown Pima cotton plants at high temperature. Plant Cell Environ. 27, 717-724.

Wu,Y., Kuzma. J., Marechal, E., Grraeff, R.,Lee, H.C., Foster, R and Chua, N.H. 1997.Abscisic acid signaling through cyclic ADP- Ribose in plants Science 278: 2126-2130.

Worral, D., Elios, L., Ashford, D., Smallwood, M., Sidebottom, C., Lillford, P., Telford, J., Holt, C. and Boweles, D. 1998. A carrot leucine-rich repeat protein that inhibits ice recrystallization. *Science* **282** : 115-116.

Xiong, L., Lee, H., Ishitani, M., Zhu, J.-K., 2002. Regulation of osmotic stress responsive gene expression by LOS6/ABA1 locus in *Arabidopsis. J. Biol. Chem.* **277**, 8588–8596.

Xu, S., Li, J., Zhang, X., Wei, H., Cui, L., 2006. Effects of heat acclimation pretreatment on changes of membrane lipid peroxidation, antioxidant metabolites, and ultrastructure of chloroplasts in two cool-season tUlfgrass species under heat stress. *Environ. Exp. Bot.* 56, 274-285.

Yamada, M., Hidaka, T., Fukamachi, H., 1996. Heat tolerance in leaves oftropical fruit crops as measured by chlorophyll fluorescence. *Sci. Hortic.* 67, 39-48.

Yamaguchi-Shinozaki, K. and Shinozaki, K. 1994. A novel *Cis*-acting element in the *Arabidopsis* gene is involved in responsiveness to drought, low temperature, or high-salt stress. Pl. Cell **6** : 251-264.

Yamane, Y., Kashino, Y., Koike, H., Satoh, K, 1998. Effects of high temperatures on the photosynthetic systems in spinach: oxygen-evolving activities, fluorescence characteristics and the denaturation process. *Photosynth. Res.* 57,51-59.

Yang, J., Sears, R.G., Gill, B.S., Paulsen, G.M., 2002. Genotypic differences in utilization of assimilate sources during maturation of wheat under chronic heat and heat shock stresses. Euphytica 125, 179-188.

Yang, KA., Lim, CJ., Hong, J.K, Park, C.Y., Cheong, Y.H., Chung, W.S., Lee, KO., Lee, S.Y., Cho, MJ., Lim, C.O., 2006. Identification of cell wall genes modified by a permissive high temperature in Chinese cabbage. *Plant Sci.* 171,175-182.

Yoshiba, Y., Kiyosue, T., Nakashina, K., Yamaguchi-Shinozaki, K. and Shinozaki, K. 1997. Regulation of levels of proline as an osmolyte in plants under water stress. *Pl. Cell* Physiol 38:1095-1102

Young, L.W., Wilen, R.W., Bonham-Smith, P.C., 2004. High temperature stress of *Brassica napus* during flowering reduces micro- and megagametophyte fertility, induces fruit abortion, and disrupts seed production. *J. Exp. Bot.* 55,485-495.

❑❑❑

Chapter 2

Response of Cyanobacteria to Ultraviolet Radiation : Mitigation Strategies to Resist the Stress

SUNITA KUMARI, RICHA, RASHUNI CHANDRA AND RAJESHWAR P. SINHA

INTRODUCTION

Cyanobacteria are a group of Gram-negative prokaryotes having higher plant-type oxygenic photosynthesis. They are cosmopolitan in distribution ranging from hot springs to Arctic and Antarctic regions. They were perhaps the first organisms to release oxygen into the then oxygen-free atmosphere. Thus, cyanobacteria were responsible for a major global evolutionary transformation leading to the development of aerobic metabolism and the subsequent rise of higher plants and animal forms. They can be used as a model organism for studying the adverse effects of UV-B because they were exposed early in their evolution to UV-B fluxes much higher than present UV-B levels (Castenholz and Garcia-Pichel, 2000).

Industrialization in the past few decades has resulted in an increase in anthropogenically released chlorofluorocarbons, chlorocarbons and organobromides causing depletion of the stratospheric ozone layer (Crutzen, 1992). Considerable amounts of reactive nitrogen species (RNS) such as nitric oxide (*NO), peroxynitrite ($ONOO^-$) and nitrous oxide (N_2O), produced naturally from unpolluted terrestrial and aquatic ecosystems or from anthropogenic sources, also contribute to the depletion of the ozone layer (Kramlich and Linak, 1994). The process of ozone depletion has been reported at mid latitudes and especially in the Antarctic where ozone levels have been reported to decline by more than 70 % during late winter and early spring in the polar vortex (Smith *et al.*, 1992). Due to ozone depletion there is an increase in the amount of harmful ultraviolet radiation on the Earth's surface which is absorbed by biomolecules such as nucleic acids and proteins ultimately resulting in lethal effects on the biological systems (Häder *et al.*, 2007).

Laboratory of Photobiology and Molecular Microbiology, Centre of Advanced Study in Botany, Banaras Hindu University, Varanasi - 221 005 U.P.

In cyanobacteria, more than 99 % of the UV-B is absorbed by chlorophyll-binding proteins and phycobilisomes, the light-harvesting complexes (Lao and Glazer, 1996). Active oxygen species is produced during intense solar radiation by transfer of excessive excitation energy from excited chlorophyll molecules to ground state (triplet) oxygen molecules. The resulting singlet oxygen is believed to damage the D1 proteins. Reactive oxygen species (ROS) is an important component in signal transduction pathway, resulting in the inhibition of gene expression of key proteins involved in photosynthesis during UV-B exposure (Mackerness *et al.*, 1999). One of the main targets of the UV-B radiation is Photosystem II (PSII), whose electron transport is inhibited and D1 and D2 reaction center proteins are degraded (Wang and Tang, 2008). RuBISCO is the primary target involved in UV-induced inhibition (Bischof *et al.*, 2002). The consequent reduction in CO_2 assimilation may be the result of both a declining activity and a reduced content of the enzyme. On contrary to the deleterious effects of UV-B, there is stimulation of nitrate reductase activity upon UV-B exposure, which might be due to accelerated uptake of NO_3^- due to changed membrane permeability or due to presence of some specific UV-B photoreceptor in the cells. UV-induced formation of thymine dimers have been reported in a number of rice-field cyanobacteria such as *Anabaena* sp., *Nostoc* sp. and *Scytonema* sp. (Sinha *et al.*, 2001a).

After all these deleterious effects, cyanobacteria still survive in the changing environmental conditions. To mitigate the deleterious effects of UV-B, certain cyanobacteria synthesize colorless, water-soluble compound, mycosporine-like amino acids (Ehling-Schulz *et al.*, 1997; Garcia-Pichel and Castenholz, 1991; Gröniger *et al.*, 2000; Sinha *et al.*, 1999a; Sinha and Häder, 2006) and lipid-soluble yellow-brown sheath pigment, scytonemin (Sinha *et al.*, 1999b). MAAs has also been characterized in Antarctic and halophilic cyanobacteria (Oren, 1997; Quesada and Vincent, 1997). Adverse effects of UV-B radiation on growth, survival, cell differentiation, motility, orientation, change in pigment composition, protein profile, photosynthesis, nitrogen metabolism and DNA, and the protective measures adopted by cyanobacteria (Fig. 1) are discussed below.

CYANOBACTERIA AND ULTRAVIOLET RADIATION

UV-B radiation has been shown to severely affect the growth and survival of several cyanobacteria (Tyagi *et al.*, 1992; Sinha *et al.*, 1995, 1996; Quesada and Vincent, 1997; Prasad and Zeeshan, 2004; Zhou *et al.*, 2009). It has been suggested that growth and survival strongly depend on the efficiency of photoreactivation or excision repair of DNA (Sinha and Häder, 1998). Various species differ with respect to their tolerance to UV-B radiation and even closely related strains show differential sensitivity. Complete killing and loss of survival within 120-180 min of UV-B exposure has been reported for several cyanobacteria (Tyagi *et al.*, 1992; Sinha *et al.*, 1995). Strains such as *Nostoc commune* and *Scytonema* sp. whose filaments are embedded in a mucilage sheath are more tolerant to UV-B radiation than the species which do not have such coverings (Sinha *et al.*, 1996). A phycoerythrin rich (brown) strain of *Nostoc* sp. was found to be more tolerant to UV-B radiation than the phycocyanin rich (blue-green) form of this strain (Tyagi *et al.*, 1992). Two Antarctica cyanobacteria, *Phormidium murrayi* and *Oscillatoria priestleyi*, isolated from a common habitat showed differential sensitivity to UV-B radiation. In contrast to complete suppression of growth in *Oscillatoria priestleyi*, only a 62 % inhibition of growth of *Phormidium murrayi* was observed following exposure to the same quantum of UV-B intensity (Quesada and Vincent, 1997). Survival and growth has been shown to be decreased with selected doses of UV-B (30 and 90 min) in non-heterocystous cyanobacterium, *Plectonema boryanum* (Prasad and Zeeshan, 2004). Han *et al.* (2003) reported inhibition of growth in a rice-field cyanobacterium *Anabaena* sp. during exposure to PAR + UVR while it was inhibited by up to 40 % by solar UVR in *Anabaena* sp. PCC 7120 (Gao *et al.*, 2007).

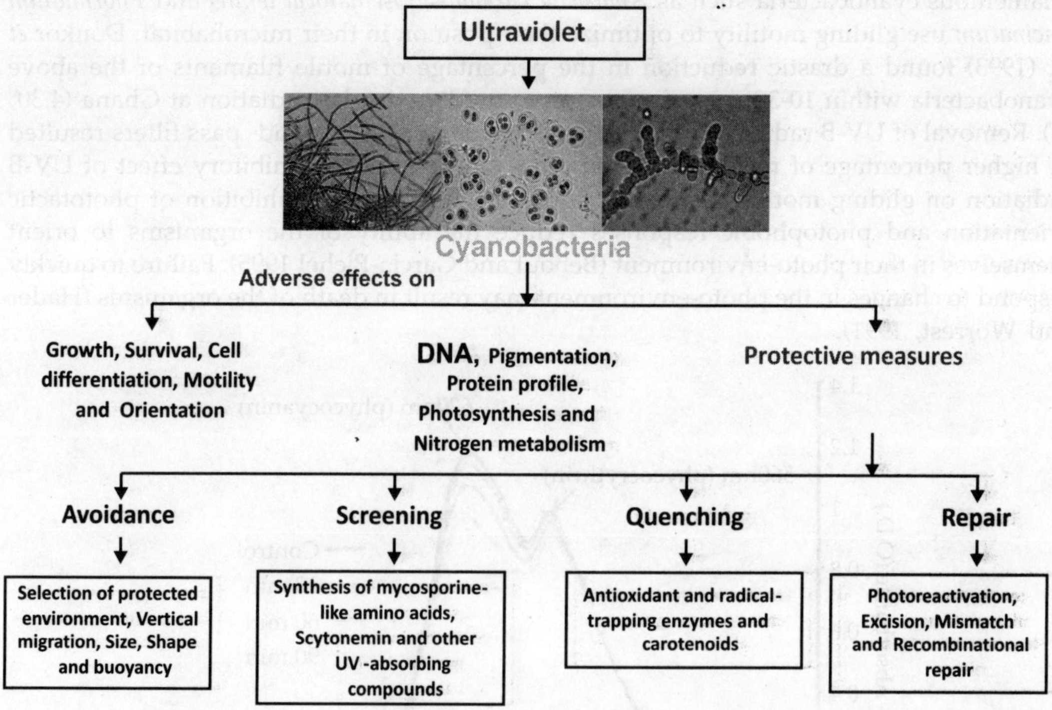

Fig. 1 : Schematic representation of the effects of UV radiation on cyanobacteria and their mitigation strategies

Three types of cell differentiations are found in the vegetative phase of cyanobacteria i.e. vegetative cells, heterocysts and akinete performing photosynthesis, nitrogen fixation and coping with stress conditions respectively. Differentiation of vegetative cells into heterocysts or akinetes is also affected by UV-B radiation. Blakefield and Harris (1994) observed delayed heterocyst differentiation and akinete formation in *Anabaena aqualis* following exposure to UV-B radiation. It is presumed that UV-B radiation alters the C:N ratio which in turn affects the spacing pattern of heterocyst in a filament (Sinha *et al.*, 1996). A decrease in the level of major heterocyst polypeptides (26, 54 and 55 kDa) following UV-B treatment has been noted which is presumed to be responsible for the disruption of multi-layered heterocyst wall, the essential component for maintaining the active form of nitrogenase. Role of UV-B radiation in the opening of specific calcium channels in the heterocyst of *Anabaena* sp. has also been demonstrated (Richter *et al.*, 1999). In *Anabaena* sp. PCC 7120 differentiation of heterocysts has also been reportedly suppressed by solar UVR (Gao *et al.*, 2007). The spiral filaments of *Arthrospira platensis* have been recently reported to be broken and compressed under solar UVR (Gao *et al.*, 2008).

Cyanobacteria employ motility and orientation mechanisms to escape harmful UV radiations and high light intensities but it has been reported that these avoidance mechanisms themselves are severely affected by UV-B radiation, resulting in a decrease or loss of their ability to escape from harmful UV conditions (Donkor *et al.*, 1993). Filamentous cyanobacteria such as *Anabaena variabilis, Oscillatoria tenuis* and *Phormidium uncinatum* use gliding motility to optimize their position in their microhabitat. Donkor *et al.* (1993) found a drastic reduction in the percentage of motile filaments of the above cyanobacteria within 10-30 min of exposure to unfiltered solar radiation at Ghana (4.30^0 N). Removal of UV-B radiation by an appropriate long-pass or band- pass filters resulted in higher percentage of motility in most cases suggesting the inhibitory effect of UV-B radiation on gliding motility of these cyanobacteria. Similarly, inhibition of phototactic orientation and photophobic responses reduce the ability of the organisms to orient themselves in their photo-environment (Bebout and Garcia-Pichel 1995). Failure to quickly respond to changes in the photo-environment may result in death of the organisms (Häder and Worrest, 1991).

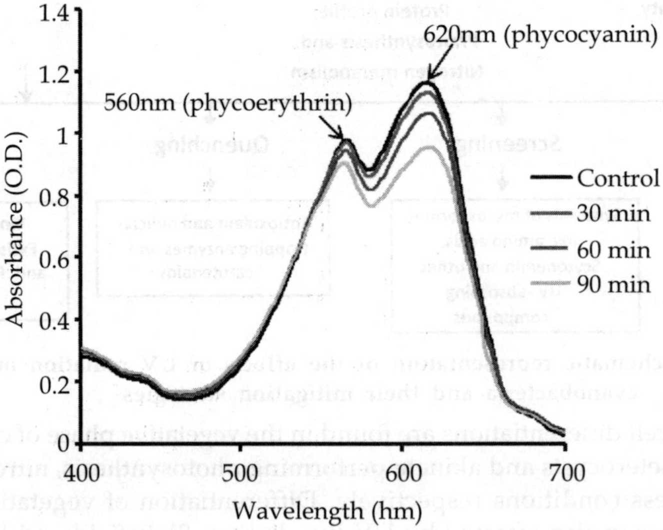

Fig. 2 : Absorption spectra showing the degradation of phycobiliproteins in *Nostoc* sp. HKAR-2 after UV-B irradiation.

It has been demonstrated that UV-B radiation causes photobleaching of photosynthetic pigments, a decrease in phycocyanin/chlorophyll *a* ratio and disintegration of phycobilisome complex leading to the bleaching of phycobiliprotein pigment complex (Häder and Worrest 1991; Tyagi *et al.*, 1992; Quesada and Vincent 1997) (Fig. 2). High PAR treatment was found to decrease the phycocyanin level by 93 %, the allophycocyanin level by 75 % and the phycoerythrin level by 69 % in *Nostoc spongiaeforme* (Bhandari and

Sharma, 2006). There are several studies indicating that UV-B radiation also influences the chlorophyll and carotenoids contents in cyanobacteria (Gao *et al.*, 2007; Han *et al.*, 2003; He and Häder, 2002a; Lesser, 2008). Phycobiliprotein seems to be the most sensitive pigment most probably due to its water soluble property. Sinha *et al.* (1995) have reported a decrease in the phycobiliprotein content and disassembly of phycobilisomal complex following UV-B irradiation in a number of cyanobacteria. The pattern of fluorescence emission spectra of phycobiliproteins observed after UV-B irradiation suggests an impairment of energy transfer from the accessory pigments to the photosynthetic reaction centers (Sinha *et al.*, 1995). Exposure of cyanobacterial cultures to intense UV-B radiation causes bleaching of phycocyanin and phycoerythrin which is in accordance with the disintegration of phycobilisomes. UV-B-induced damage of photosynthetic pigments, phycocyanin, carotenoids and chlorophyll *a* has been reported but phycocyanin appeared to be the main target of UV-B in *Plectonema boryanum* (Prasad and Zeeshan, 2004). UV-B radiation also induces the lipid peroxidation of polyunsaturated fatty acids (PUFA) via oxidative damage which subsequently affects the integrity of cellular and thylakoid membranes (He and Häder, 2002b).

Photosynthetic parameters such as $^{14}CO_2$ uptake, O_2 evolution and ribulose-1,5 bisphosphate carboxylose (RuBISCO) activity are greatly inhibited following exposure to UV-B radiation (Wilson *et al.*, 1995; Quesada and Vincent, 1997). Photosynthetic O_2 evolution of a *Phormidium* strain from Lake Baikal declined within minutes of exposure to solar irradiation (Häder and Worrest, 1991) while Gerber and Häder (1995) had reported reduction in photosynthetic O_2 evolution in five species of phytoplankton after UV-B exposure. Several workers have reported an inhibition of $^{14}CO_2$ uptake by a number of cyanobacteria following exposure to UV-B irradiation (Newton *et al.*, 1979; Tyagi *et al.*, 1992; Sinha *et al.*, 1996). In a study of diurnal variability of the effect of solar UV-radiation on photosynthesis in an algal mat comprising of *Lyngbya aestuarii*, Cockell and Rothschild (1999) found a mean increase in the total primary productivity (measured in terms of $^{14}CO_2$ uptake) over the day when solar UV-B was screened out. In *Plectonema boryanum*, ^{14}C-fixation was found to be more sensitive to UV-B than photosynthetic oxygen evolution. Photosynthetic electron transport activity decreased in response to UV-B. This inhibition was due to a greater susceptibility of PSII than PSI (Prasad and Zeeshan, 2004). UV radiation inhibits photosynthesis of desert *Nostoc* sp. The effect was dependent on the radiation intensity and exposure time. Exogenous chemicals such as NAC (N-acetylcysteine), ASC (ascorbate acid), and SNP (sodium nitroprusside) evidenced different levels of protection of photosynthesis activity under UV radiation. NAC addition can provide effective protection for photosynthetic activity of *Nostoc* under UV-B radiation. Protective effect of NAC has been shown to be greater than those of ASC and SNP addition after 2 h of UV treatment (Wang *et al.*, 2008). It has been shown that RuBISCO may undergo a number of modifications such as photo degradation, fragmentation and denaturation of polypeptide chain, change in active site and increased solubility of membrane proteins. In addition, supply of ATP and $NADPH_2$ may also be inhibited which may cause inhibition of CO_2- fixing ability. The D_1 and D_2 polypeptides that are the major constituents of PS-II reaction centers are also degraded, even after exposure to intermediate levels of UV-B radiation (Sass *et al.*, 1997; Campbell *et al.*, 1998). The D1 core protein of photosystem

II is highly susceptible to damage by UV-B-generated radicals. Two genes, *psbA2* and *psbA3,* encoding identical D1 polypeptides were induced four to six fold by UV-B irradiation at 20 and 60 microeinsteins $m^{-2} s^{-1}$ and intense white light (Mate *et al.*, 1998). Down-regulation of photosynthetic genes (except *psbA*) suggested a decrease of photosynthetic activity as UV-B exposure intensified from 20 to 60 microeinsteins $m^{-2} s^{-1}$ (Huang *et al.*, 2002). UV-B radiation may also induce modifications and photo degradation of membrane proteins, which may subsequently affect the integrity of thylakoid membrane thereby causing the damage of photosynthetic components which are other wise essential for active CO_2 fixation.

The nitrogen-fixing enzyme nitrogenase has been found to be extremely sensitive to UV-B radiation (Newton *et al.*, 1979; Tyagi *et al.*, 1992; Kumar *et al.*, 1996; 2003; Lesser, 2008). In a *Nostoc* sp., nitrogenase activity was lost within 45 min of UV-B exposure whereas nitrate reductase and glutamine synthetase activity remained more or less unaffected (Tyagi *et al.*, 1992). Kumar *et al.* (2003) reported complete inactivation of nitrogenase activity within 25-40 min of UV-B exposure in a number of cyanobacteria including *Anabaena* sp., *Nostoc* sp., *Calothrix* sp. and *Scytonema* sp. Employing inhibitors of protein synthesis and PS-II activity they have demonstrated that restoration of nitrogenase activity in UV-B-treated cultures occurs only after fresh synthesis of nitrogenase polypeptide. The inactivation of activity most probably occurs due to complete damage to the nitrogenase polypeptide. It has been suggested that the aromatic amino acids present or the native structure of nitrogenase are responsible for extreme susceptibility of nitrogenase to UV-B radiation.

UV-B radiation is known to cause a subtle change in the protein profile of several cyanobacteria (Sinha *et al.*, 1995; Sinha and Häder, 2002a). In a *Nostoc* sp., the low molecular weight proteins of around 20 kDa (ab monomers of phycocyanin) were the most affected one. It is interesting to find that UV-B radiation has a drastic effect specifically on low molecular weight proteins. In a comparative study on total protein profile of several cyanobacteria, Sinha *et al.* (1995), reported complete loss of protein bands between 14.5 and 45 kDa after 90 and 120 min of UV-B exposure in *Nostoc carmium* and *Anabaena* sp. However protein bands of relatively higher molecular weight viz., 55 and 66 kDa were unaffected even after 120 min. In species such as *N. commune* and *Scytonema* sp., which contain a mucilage covering/sheath, complete elimination of bands occurred only after 150 min of UV-B exposure (Sinha *et al.*, 1995). The increased tolerance to UV-B radiation in certain species may be ascribed to the presence of mucilage covering or photoprotectants such as scytonemin and mycosporine-like amino acids (MAAs) in the cell. Appearance of certain new proteins may be due to UV-B shock and may be similar to the *hsp* group of proteins (Shibata *et al.*, 1991). In *Nostoc commune*, semiquantitative analysis of about 1350 proteins revealed that at least 493 proteins (37 %) belong to the UV-B stimulon. A minimum of 188 proteins were positively stimulated, whereas a minimum of 305 proteins were negatively affected (Ehling-Schulz *et al.*, 2002). Translation activity was also suppressed by oxidative stress in *Synechocystis* sp. PCC 6803 and oxidation of the elongation factor EF-G was found to be responsible for the inhibited translation activity (Kojima *et al.*, 2007).

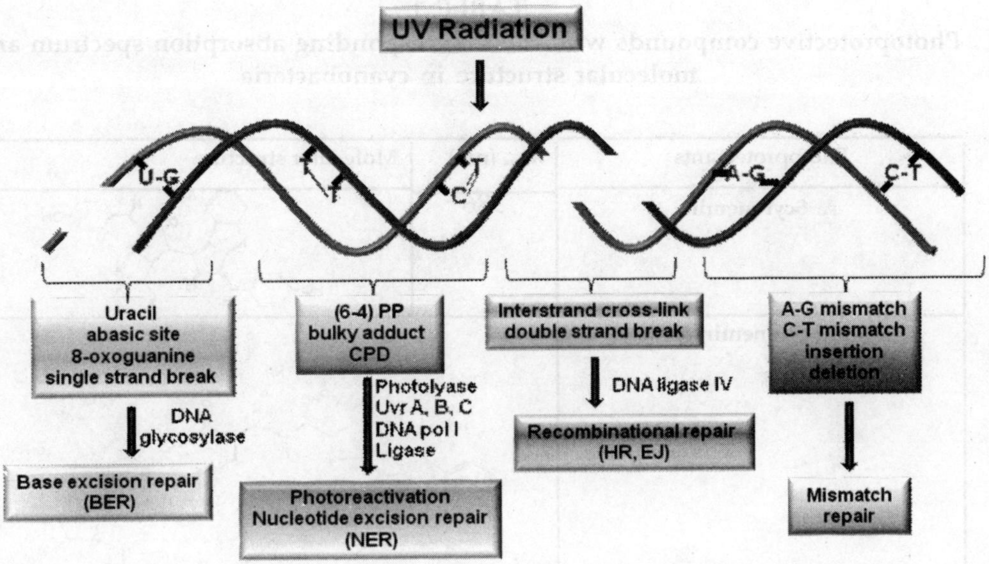

Fig. 3 : UV-induced DNA damage and their repair mechanisms

UV-B-induced lesions in DNA have been reported. These lesions result either due to direct absorption of UV-B radiation by DNA or from free radicals and reactive oxygen species (ROS) formed by various photochemical processes (Fig. 3). The main class of UV-induced lesions consist of dimeric photoproducts such as cis-syn cyclobutane pyrimidine dimers (CPDs) and pyrimidine (6-4) pyrimidone photoproducts (6-4 PPs) and their Dewar isomers (Mitchell and Karentz, 1993; Friedberg *et al.*,1995; Ravanat *et al.*, 2001; Cadet *et al.*, 2005). It has been reported that DNA lesions, if not repaired, may inhibit the progress of DNA polymerase, and interferes with DNA synthesis and RNA transcription, ultimately leading to mutation or death of the organism. (Britt, 1995; Lindahl, 1993). UV-B radiation in cyanobacteria such as proteins and DNA which have absorption maxima in this region while UV-A irradiation has indirect effects via energy transfer from UV-A-stimulated chromophores to the DNA target or via photosensitized (chlorophylls and phycobilins can act as photosensitizers) production of ROS (Hargreaves *et al.*, ?007; He and Häder, 2002b). He and Häder (2002a, 2002c) reported UV-B and UV-A-induced ROS production for the first time in cyanobacteria using a ROS-sensitive probe, 2′,7′-dichlorodihydrofluorescein diacetate (DCFH-DA). The damaging effects of UV-B radiation on DNA and formation of various photoproducts are well known in a number of organisms including the Antarctic marine planktonic diatom communities (Karentz *et al.*, 1991a; Buma *et al.*, 1995). However, little information is available regarding the specific effects of UV-B radiation on DNA of cyanobacteria, although few studies have been conducted with UV-C radiation (Levine and Thiel, 1987; Eker *et al.*, 1990). Sinha *et al.* (2001a) have demonstrated that exposure to UV-B radiation could induce the formation of thymine dimers in three rice-field cyanobacteria, *Nostoc* sp., *Anabaena* sp. and *Scytonema* sp. By using dot-blot and chemiluminescence method these workers successfully developed a relatively quick, simple and efficient method for the quantitative analysis of thymine

TABLE 1
Photoprotective compounds with their corresponding absorption spectrum and molecular structure in cyanobacteria

Photoprotectants	λ_{max} (nm)	Molecular structure
A. Scytonemin	386	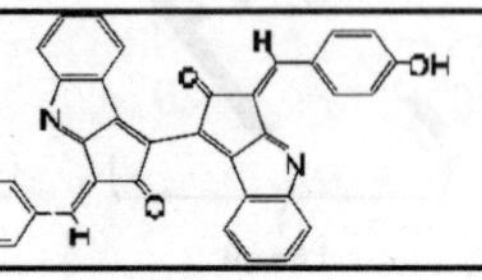
B. Scytonemin A		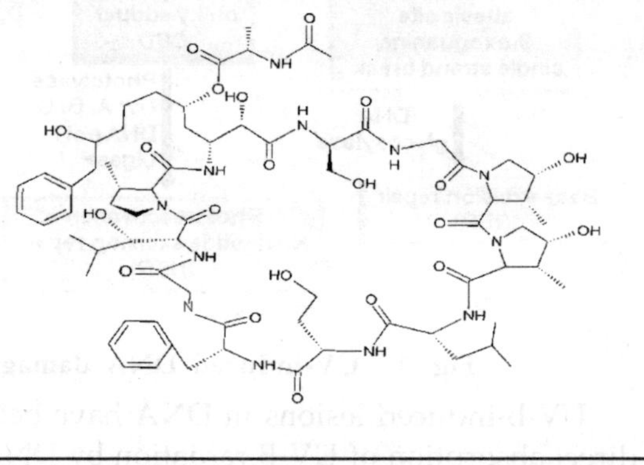
C. Mycosporine-like amino acids		
Palythinol	332	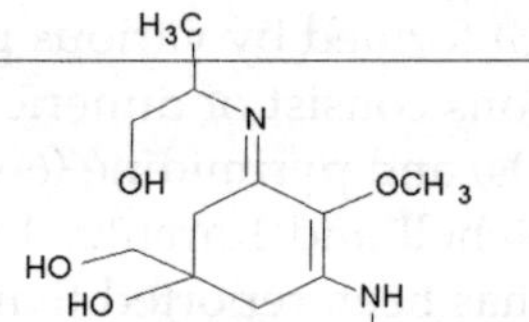
Shinorine	334	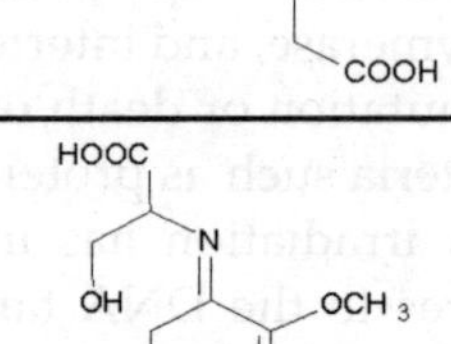
Porphyra-334	334	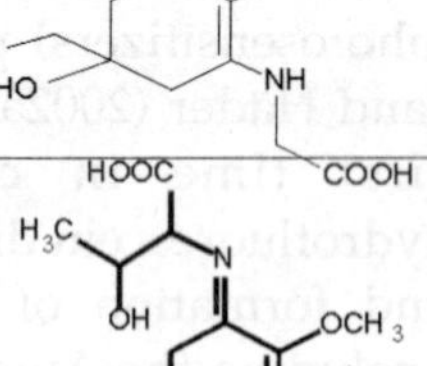
Palythene	360	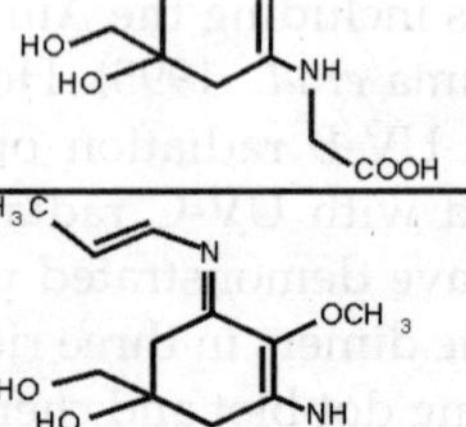

dimers in aquatic primary producers including cyanobacteria (Sinha *et al.*, 2001a). Readers interested in detailed information about the DNA damage and commonly used techniques for their detection are referred to the review articles by Sinha and Häder (2002b) and Kumari *et al.* (2008). Kumar *et al.* (2004) observed a hypochromic effect in genomic DNA of the *Anabaena* strain BT2 after UVR exposure and reported damaged template activity of DNA following UVR exposure using a PCR-based assay such as RAPD, rDNA amplification and ARDA. A fluorimetric analysis of DNA unwinding (FADU) to quantify the DNA strand breaks induced by ROS under UV-A and UV-B exposure in the cyanobacterium *Anabaena* sp (He *et al.* 2002). Their findings suggested that both UV-A and UV-B radiation results in an evident oxidative stress and enhances the double-strand breaks in the genomic DNA. Gao *et al.* (2008) have reported temperature and biomass dependent CPDs formation under UV stress in the commercially important cyanobacterium *A. platensis*.

CYANOBACTERIA AND MITIGATION STRATEGIES

Since cyanobacteria are believed to have originated in the Precambrian era at a time when ozone shield was absent, they presumably faced high fluxes of UV radiation which must have acted as an evolutionary pressure leading to selection for efficient UV radiation protecting mechanisms. Phoenix *et al.* (2006) studied on a system analogous to one in the early Precambrian environment and they found cyanobacterial mats encased in amorphous silica matrices were protected from lethal UVR and proposed that these deposits would have strongly attenuated Archean and early Precambrian levels of UVR. Wu *et al.* (2005) proposed that decreased helix pitch in presence of UV-B to obtain a more compact structure of the spirals, which eventually results in self-shading, is an effective protective mechanism against photoinhibition in *Arthrospira platensis*. Tolerance of cyanobacteria to intense sunlight as well as UV radiation might have contributed to their success during early stages of colonization. As a consequence they have developed effective mechanisms to counteract the damaging effects of UV-B radiation. In case of microbes behavioral responses as well as external and internal mechanisms of protection against UV-induced damages have been reported and it has been proposed that several of these mechanisms for UV-protection and/or repair of UV-induced damage might have worked singly or in various combinations (Pierson *et al.*, 1993; Sinha and Häder, 2008). An important mechanism to prevent UV-induced photodamage, is the synthesis of UV-absorbing compounds (Singh *et al.*, 2008a).

The avoidance of damaging UVR provides protection to some extent in cyanobacteria as this mechanism depends on the motility and mat forming ability of cyanobacteria as well as on the turbidity and depth of the water column. Once UVR reaches the inside the cell, it interacts with oxygen and other organic compounds to produce toxic ROS such as superoxide (O_2^-), hydroxyl radical (OH^-) or hydrogen peroxide (H_2O_2) finally resulting in oxidative stress. To overcome the oxidative stress cyanobacteria have developed an antioxidant system as a second line of defense mechanisms against UVR. This system includes non-enzymatic and enzymatic antioxidants. Non-enzymatic antioxidants comprise ascorbate (vitamin C), a-tocopherol (vitamin E), carotenoids and reduced glutathione. The enzymatic antioxidants include superoxide dismutase (SOD), catalase, glutathione peroxidase (GSH-Px) and the enzymes involved in the ascorbate-glutathione cycle to detoxify ROS, such as ascorbate peroxidase (APX), monodehydroascorbate reductase (MDHAR), dehydroascorbate reductase (DHAR) and glutathione reductase (GR) (He and Häder, 2002b). Carotenoids protect cells against photooxidative damage by absorbing triplet state energy from chlorophyll and quenching singlet state oxygen while á-tocopherol prevents lipid peroxidation by scavenging ROS (Niyogi, 1999). However, ascorbate performs direct quenching of ROS, regenerate a-tocopherol and acts as a substrate in both violaxanthin de-epoxidase and APX reactions. Glutathione protects thiol groups in various enzymes and is also involved in a-tocopherol and ascorbate regeneration through the glutathione-ascorbate cycle (Niyogi, 1999). SOD scavenges superoxide radicals and converts them to hydrogen peroxide which is further converted to water and O_2 via a combined catalase-peroxide system (Tel-Or *et al.*, 1986). In the cyanobacterium *N. commune*, changes in the carotenoid pattern in response to UV-B irradiation was reported and myxoxanthophyll and echinenone were suggested to act as outer membrane-bound UV-B photoprotectors (Ehling-Schulz *et al.*, 1997). Rubrerythrin-homologue (RbrA) (peroxidase), common in air-sensitive bacteria was identified in *Anabaena* PCC 7120 and was found to protect the nitrogenase enzyme against oxidative stress (Zhao *et al.*, 2007). In addition, cyanobacteria synthesize photoprotectants such as scytonemin and mycosporine-like amino acids (MAAs), having strong absorption in the UV-A and/or UV-B region of the spectrum play an important role in allowing these organisms to grow and survive in habitats exposed to strong irradiation (Sinha and Häder, 2008). In a terrestrial *Nostoc commune* strain, subjected to regular cycles of desiccation and rewetting, UV-absorbing compounds play a key role as UV-protectants (Ehling-Schulz *et al.*, 1997; Oren and Gunde-Cimerman, 2007).

Scytonemin is a yellow-brown colored pigment occurring in the extracellular sheath of a number of terrestrial cyanobacteria especially in those species growing in habitats exposed to intense solar radiation (Garcia-Pichel and Castenholz, 1991; 1993). It is a lipid soluble compound having *in vivo* absorption maximum at 370 nm in the UV-A region but shows considerable absorption in the UV-B region also. Purified scytonemin shows absorption maximum at 386 nm although there being appreciable absorbance at 252, 278 and 300 nm (Sinha *et al.*, 1999a). It is a novel dimeric pigment composed of indole and phenolic subunits with a molecular mass of 544 Da (Table 1A). Micro-spectrophotometric measurements of the transmittance of pigmented sheath and the quenching of ultraviolet excitation of phycocyanin fluorescence suggest that the pigment is effective in shielding the cells from incoming UV radiation and is able to reduce the incident UV-A radiation entering the cells to around 90 % (Garcia-Pichel and Castenholz, 1991; Brenowitz and Castenholz, 1997). Scytonemin may constitute as much as 5 % of the total cellular dry weight or even higher in some naturally occurring cyanobacteria (Castenholz, 1997). Since rapid photodegradation of scytonemin does not occur, it is highly stable and able to persist very long in terrestrial cyanobacterial crusts or dried mats (Brenowitz and Castenholz, 1997; Quesada *et al.*, 1999). It has been proposed that appearance of scytonemin during the Precambrian era must have helped the cyanobacteria in the colonization of exposed, shallow water and terrestrial habitats (Quesda *et al.*, 1999). Besides having a role as UV protectant (Fleming and Castenholz, 2007), scytonemin may have additional roles such as protection against pathogenic attack, bacterial decomposition or herbivore grazing

(Sinha and Häder, 2002a; 2008). Recently, scytonemin was extracted from *Lyngbya* sp. (inhabiting rocks at the shore near Sao Fransisco do Sul, Brasil) in 100 % acetone and the organism was recultivated and again after three weeks, scytonemin was extracted (Richter *et al.*, 2006). A gene cluster (NpR1276–NpR1259) responsible for scytonemin biosynthesis was identified in the genome of *N. punctiforme* ATCC 29133 by random transposon mutation (Soule *et al.*, 2007). This 18-gene cluster contains genes involved in the biosynthesis of tryptophan and tyrosine while other genes do not show any homology with functionally characterized proteins. Recently, Balskus and Walsh (2008) have presented the possible biosynthetic route for the scytonemin biosynthesis and identified the acyloin reaction as a key step in constructing the carbon framework of this ecologically and evolutionary important pigment. This pigment has been found to act as ultraviolet sunscreen, with the greatest absorption in the spectral range of UV-A (Proteau *et al.*, 1993) and therefore may have application in sunscreens. Production of this pigment in certain cyanobacteria is believed to be the earliest developed mechanism of ultraviolet protection, more ancient than the flavonoids or melanins (Garcia-Pichel, 1998). Its ring structure, the "scytoneman skeleton", is unique among natural products and is thought to stem from the condensation of tryptophan- and phenylpropanoid-derived subunits and also closely related to nostodione A (Proteau *et al.*, 1993). Other attractive structural features include its lack of chirality, multiple dissection points, and phenolic groups that could be easily modified. These attributes and its relation to other antiproliferative agents make scytonemin a prime candidate for investigating its potential utility as a pharmacophore with which new therapies targeting hyperproliferative disorders can be developed. The cyclic peptide, scytonemin A (Table 1B), from a *Scytonema* sp. has also been shown to be a strong calcium agonist (Helms *et al.*, 1988). In a screening effort designed to look for inhibitors of a cell cycle kinase human *polo*-like kinase, a serine/threonine kinase that plays an integral role in regulating the G2/M transition in the cell cycle, scytonemin was shown to be active, with an IC_{50} of 2 mM (Stevenson *et al.*, 2002a). Scytonemin also inhibits other cell cycle kinases with similar potency. In human T-cell leukemia Jurkat cells, scytonemin inhibited cell proliferation (IC_{50} = 7.8 mM) and induced apoptosis in 24% of the cells (at 3 mM). Scytonemin inhibited *polo*-like kinase 1 activity in a concentration-dependent manner with an IC_{50} of 2 mM against the recombinant enzyme. Biochemical analysis showed that scytonemin reduced GST-*polo*-like kinase 1 activity in a time-independent fashion, suggesting reversibility, and with a mixed-competition mechanism with respect to ATP. Although scytonemin was less potent against protein kinase A and Tie2, a tyrosine kinase, it did inhibit other cell cycle-regulatory kinases like Myt1, checkpoint kinase 1, cyclin-dependent kinase 1/cyclin B, and protein kinase Câ2 with IC_{50} values similar to that seen for *polo*-like kinase 1. Consistent with these effects, scytonemin effectively attenuated, without chemical toxicity, the growth factor- or mitogen-induced proliferation of three cell types commonly implicated in inflammatory hyperproliferation. Similarly, scytonemin (up to 10 mM) was not cytotoxic to nonproliferating endotoxin-stimulated human monocytes. In addition, Jurkat T cells treated with scytonemin were induced to undergo apoptosis in a non-cell cycle-dependent manner consistent with its activities on multiple kinases. Scytonemin possesses both anti-inflammatory and anti-proliferative properties. The dual kinase inhibitory activity may be of value therapeutically in acute and possibly chronic disorders where both inflammation and proliferation are prevalent. Limiting both neovascularization and the presence of inflammatory mediators at the affected site offers a broader spectrum of activity, which could hypothetically be more efficacious in the management of complex inflammatory disorders. Whether or not the ability of scytonemin to inhibit both inflammation and proliferation is due solely to its effects on these two kinases, scytonemin offers a novel pharmacophore, which may serve as a template for synthesizing more potent and selective inhibitors (Stevenson *et al.*, 2002b).

The other class of photoprotectants is colorless, water-soluble MAAs having absorption maxima in the UV range between 310 - 362 nm (Torres *et al.*, 2006; Callone *et al.*, 2006; Conde *et al.*, 2007; Sinha *et al.* 2007; Singh *et al.*, 2008a). First identified in fungi, these compounds are now reported to be present in a number of taxonomically diverse groups of organisms from eukaryotic and prokaryotic classes. They are present in cyanobacteria (Garcia-Pichel *et al.*, 1993; Sinha *et al.*, 1999b; 2001b), eukaryotic algae (Carreto *et al.*, 1990; Karentz *et al.*, 1991b), marine heterotrophic bacteria and marine invertebrates (Karentz *et al.*, 1991b), fish (Dunlap *et al.*, 1989) and in a wide variety of other marine organisms (Karentz *et al.*, 1991b; McClintock and Karentz, 1997; Dunlap and Shick, 1998). Among cyanobacteria, MAAs has been reported from species growing in freshwater, marine or terrestrial habitats (Garcia-Pichel and Castenholz, 1993; Garcia-Pichel *et al.*, 1993; Zhang *et al.*, 2007). Their presence has also been reported from strains growing in Antarctica (Quesada and Vincent, 1997) as well as from a community of halophilic species (Oren, 1997).

The synthesis of MAAs occurs in bacteria, cyanobacteria, phytoplankton and macroalgae but not in animals, because they lack the shikimate pathway. Recently, the gene for the shikimate pathway in the metazoan *Nematostella vectensis* has been disclosed by Starcevic *et al.* (2008). These genes were transferred from cyanobacteria to dinoflagellates and finally to metazoans by lateral gene transfer events. Structurally these compounds contain cyclohexenone or cyclohexenimine chromophore conjugated with the nitrogen substituent of an amino acid or its imino alcohol and have an average molecular weight of 300 (Cockell and Knowland, 1999). A number of MAAs such as shinorine, mycosporine-glycine, palythene, palythinol, asterina-330 and porphyra-334 (Table 1C) has been reported in cyanobacteria (Sinha *et al.*, 2007). Recent report suggests that the MAA shinorine in *Anabaena variabilis* PCC 7937 is synthesized under various abiotic stressors with or without UVR (Singh *et al.*, 2008b), indicating their additional role. The synthesis of MAAs in cyanobacteria is also dependent on available nitrogen (Singh *et al.*, 2008b), and growth media with a nitrogen source support the highest MAA synthesis in comparison to others without a nitrogen source (Singh *et al.*, 2008c). However, recently, the biosynthesis of three MAAs (mycosporine-glycine, porphyra-334 and shinorine) in the rice-field cyanobacterium *A. doliolum* was reported. Mycosporine-glycine was found to be under constitutive control and other two MAAs were specifically induced by UV-B radiation (Singh *et al.*, 2008d). This indicates that different strains of *Anabaena* differ from each other in the expression of genes that are involved in the conversion of the primary MAA to secondary MAAs. Recently, Oyamada *et al.* (2008) reported the photoprotection of human fibroblast cells against UVR and their proliferation by three MAAs (shinorine, porphyra-334 and mycosporine-glycine). The specific MAA content in several isolates has been found to vary between 0.16 to 0.84 % of the total dry weight. Sinha *et al.* (2001b) have demonstrated the induction of MAAs by solar ultraviolet-B radiation in three cyanobacteria namely, *Anabaena* sp., *Scytonema* sp. and *Nostoc commune*. They reported that PAR and UV-A were ineffective in causing induction of MAAs. All these cyanobacteria possessed only a single type of MAA which was identified as shinorine. Zhang *et al.* (2007) isolated and characterized three MAAs, M-tau (UV-B sunscreen), dehydroxylusujirene and M-343 in *Synechocystis* sp. PCC 6803 and observed that dehydroxylusujirene and M-343 induction was mainly UV-A dependent but M-tau could be induced by both UV-A and UV-B.

It has been suggested that the content of MAAs in the cell is regulated by osmotic mechanisms (Oren, 1997), may transfer energy to the photosynthetic reaction centers (Sinha *et al.*, 1999b), may have a role in osmotic regulation (Oren, 1997) and, they may act as antioxidants and prevent cellular damage resulting from UV-induced production of highly reactive oxygen species. Interestingly, in laboratory-grown cultures of *Chlorogloeopsis* sp., both UV and osmotic stress are able to induce and regulate the synthesis of MAAs (Portwich and Garcia-Pichel, 1999; 2000). In this species, a specific UV-B photoreceptor, probably pterin may mediate MAA induction (Portwich and Garcia-Pichel, 2000). Misonou *et al.* (2003) have found that MAAs effectively block thymine dimer formation by UVR in vitro, however, *in vivo* experiments are needed to prove the role of MAAs in protecting DNA of organisms.

A terrestrial species of *Nostoc commune* when exposed to high solar radiation was found to possess a UV-A/B absorbing pigment with absorption maxima at 312 and 325 nm (Scherer *et al.*, 1988). In this species, an oligosaccharide MAA (OS-MAA), has been shown to be located in the extracellular glycan sheath (Böhm *et al.*, 1995; Ehling-Schulz *et al.*, 1997). It is the first mycosporine so far known to be covalently linked to the oligosaccharide and located outside the cell (Böhm *et al.*, 1995). This pigment besides providing protection by absorbing UV radiation may also provide additional protection by quenching (Dunlap and Yamamoto, 1995). The UV-absorbing compounds, MAAs are true 'multipurpose' secondary metabolites considered as natural photoprotectants that may be biotechnologically exploited in various ways. The UV-A absorption properties of the crude methanolic extract of Porphyra-334 have been determined against two commercial sun care products in terms of mean critical wavelength, mean UV-A/UV-B ratios and UV-A protection category (Torres *et al.*, 2006) and data obtain suggest that it can serve as a UV-A protecting sunscreen by providing wide protection against UV radiation. However to fulfill the sunscreen criteria the amino acids or amino alcohol functions in the MAAs has to be replace by alkyl amino groups in order to reduce their hydrophilic properties. It has also been reported that MAAs may act as antioxidants to prevent cellular damage resulting from UV-induced production of active oxygen species (Dunlap and Yamamoto, 1995). Recently, Oyamada *et al.* (2008) has investigated the protective effects of MAAs (such as shinorine, porphyra-334 and mycosporine-glycine) on human fibroblast cells from UV irradiation and it was observed that all the three examined MAAs; shinorine, porphyra-334 (P-334) and mycosporine-glycine protect the cells from UV-induced cell death. However, mycosporine-glycine was found to be more effective. MAAs has potential applications in cosmetics and toiletries as a UV protectors and activators of cell proliferation. A mixture of MAAs (i.e, P-334 + shinorine) extracted from the red alga *Porphyra umbilicalis* has been reported to suppress UV-induced aging in human skin

(Daniel *et al.*, 2004). Misonou *et al.* (2003) has observed that MAAs can block the production of both 6-4 photoproduct and cyclobutane pyrimidine dimer (CPD) formation. MAAs may also act as antioxidants, preventing cellular damage resulting from UV-induced production of reactive oxygen species (ROS) (Suh *et al.*, 2003; Oren and Gunde-Cimerman, 2007). These compounds may be of great interest in the development of artificial human sunscreens and an important field for future biotechnological research. Yuan *et al.* (2009) suggested that the similar ORAC (oxygen radical absorbance capacity) activities of the grade 1 and 2 dulse extracts likely reflected the efficacy of the MAAs shinorine and usujirene as quenchers of AAPH (2,2'-azo-bis(2 amidinopropane)dihydrochloride)-derived peroxyl radicals. The greater antiproliferative effects of the grade 2 dulse extract likely reflects the absorption and bioactivity of the least polar MAA usujirene into murine skin melanoma cell membranes. Yoshiki *et al.* (2009) evaluated the antioxidant activity (diphenylpicrylhydrazyl (DPPH)-radical scavenging activity) of the low-molecular weight fraction of water-soluble Susabinori (*Porphyra yezoensis*) extract. They found for the first time an antioxidant from heat treated Nori named as NAC (Nori antioxidant compound), a dehydrated porphyra-334. Further work is needed to test the potentialities of MAAs as antioxidants and thereby opening the way for industrial or pharmacological development of the biological sunscreens and antioxidants.

Besides scytonemin and MAAs, occurrence of a few other UV-absorbing compounds, conferring resistance to UV radiation to the organisms, have been reported (Kumar *et al.*, 1996). The shielding role of a brown-coloured pigment (having major absorption peak at 314 nm) from *Scytonema hofmanii* and a phycoerythrin rich pink extract from *Nostoc spongiaeforme*, against UV-B-induced damage, has been reported (Kumar *et al.*, 1996). A strain of *Nostoc* sp. showing resistance to high light intensity and UV radiation has been reported to produce three different UV-absorbing compounds having absorption maxima at 256, 314 and 400 nm (de Chazal and Smith, 1994).

During the last two decades vast information on UV-B absorbing/sun-screening pigments from diverse organisms have been gathered and this has led to the development of a database on photoprotective compounds (Gröniger *et al.*, 2000; Sinha *et al.*, 2007). Readers may find detailed information on the absorption maxima and extinction coefficients of different compounds together with experimental protocols on the web site (see http://www. biologie.uni-erlangen.de/botanikI/index.html).

To conserve the genome integrity, nature has provided several repair pathways for proper functioning and survival of all organisms. DNA repair is a highly complex phenomena, mainly depends on the types and location of lesions in the genome within the cell (Essers *et al.*, 2006). The process of DNA repair proceeds *via* certain specialized repair proteins that scan the genome continuously and encounter the DNA lesions by triggering different repair mechanisms such as photoreactivation, base excision repair (BER), nucleotide excision repair (NER), mismatch repair (MMR), recombinational repair (RER), and some specialized forms of repair system (Carell and Epple, 1998; Sinha and Häder, 2002b) (Fig. 3). It has been demonstrated that cyanobacteria in general are very sensitive to UV-B radiation and death is inevitable in the presence of increased level of UV-B flux. Impact of ultraviolet-B radiation in causing the damages to the DNA of the cyanobacterium, *Anabaena* strain BT2 and *Nostoc muscorum* has been investigated (Kumar *et al.*, 2004). Cyanobacteria protect themselves from photodamage by adopting a number of repair mechanisms such as photoreactivation and light-independent NER (Sinha and Hader, 2001). In *Microcystis aeruginosa* H_2O_2 treatment induces PCD in this organism, and catalase was capable of inhibiting PCD, implicating the role of PCD under oxidative stress (Ross *et al.*, 2006). Several genes encoding for caspases, enzymes involved in PCD in eukaryotes, have also been recently reported in the sequenced genome of *M. aeruginosa* (Frangeul *et al.*, 2008) The caspase activity as well as proteins reacting to human caspase-3 antibodies was reported in *Trichodesmium* sp. (Berman-Frank *et al.*, 2004). The freshwater cyanobacterium *Anabaena* has also been found to activate PCD and increases general protease activity after exposure to univalent-cation salts (Ning *et al.*, 2002).

CONCLUSIONS

UV-B radiation has now been unequivocally demonstrated to show detrimental effect on a number of vital physiological and biochemical processes of cyanobacteria. UV-B affects pigmentation, disintegrates phycobilisomes, severely inhibits photosynthetic processes and causes inactivation of nitrogenase. Major alterations in general protein profile have also been reported. Formation of dimers between two adjacent pyrimidine bases, *cis*-syn cyclobutane dimers and pyrimidine (6-4) pyrimidone products following UV-B exposure in certain cyanobacteria has also been demonstrated. Till date most of the studies pertaining to UV-B radiation effects on cyanobacteria are based on short term laboratory experiments employing high UV-B flux. Only a few short-term experiments that utilize solar UV-B radiation obtained by using various long-pass filters are available (Donkor *et al.* 1993). Conclusions about the long-term ecological impacts of ambient natural UV-B radiation cannot be accurately forecasted from results obtained through short-term incubation studies. The fragmentation of cyanobacterial filaments under UVR may possibly occur due to PCD of severely damaged cells in a filament, however, degradation of proteins that maintains the structure, has been proposed for the filament breakage (Wu *et al.*, 2005).

Other environmental and biological variables should be taken into account while attempting a realistic assessment of the effects of enhanced levels of UV-B radiation. The existence of various protective/repair strategies provide a high level of natural defense against the potential effects of elevated UV-B radiation. Information related to the ecological significance of sun-screening substances such as scytonemin and MAAs and spatial distribution of MAAs from a wide range of organisms are added every day and data available suggest their great potential. There is need to explore in detail the protective role of various sunscreen compounds against UV-B radiation. Efforts are also needed to reveal the genetic level of tolerance in cyanobacteria.

REFERENCES

Balskus, E. P. and Walsh, C. T. 2008. Investigating the initial steps in the biosynthesis of cyanobacterial sunscreen scytonemin. *Journal of the American Chemical Society* 130: 15260-15261.

Bebout, B. M. and Garcia-Pichel, F. 1995. UV-B induced vertical migrations of cyanobacteria in a microbial mat. *Applied and Environmental Microbiology* 61: 4215-4222.

Berman-Frank, I., Bidle, K. D., Haramaty, L. and Falkowski, P. G. 2004. The demise of the marine cyanobacterium, *Trichodesmium* spp., via an autocatalyzed cell death pathway. *Limnology and Oceanography* 49: 997-1005.

Bhandari, R. and Sharma, P. K. 2006. High-light-induced changes on photosynthesis, pigments, sugars, lipids and antioxidant enzymes in freshwater (*Nostoc spongiaeforme*) and marine (*Phormidium corium*) cyanobacteria. *Photochemistry and Photobiology* 82: 702-710.

Bischof, K., Kräbs, G., Wiencke, C. and Hanelt, D. 2002. Solar ultraviolet radiation affects the activity of ribulose-1,5-biphosphate carboxylase-oxygenase and the composition of photosynthetic and xanthophyll cycle pigments in intertidal green alga *Ulva lactuca* L. *Planta* 215: 502-509.

Blakefield, M. K. and Harris, D. O. 1994. Delay of cell differentiation in *Anabaena aequalis* caused by UV-B radiation and the role of photoreactivation and excision repair. *Photochemistry and Photobiology* 59: 204-208.

Böhm, G. A., Pfleiderer, W., Böger, P. and Scherer, S. 1995. Structure of a novel oligosaccharide-mycosporine-amino acid ultraviolet-A/B sunscreen pigment from the terrestrial cyanobacterium *Nostoc commune. Journal of Biological Chemistry* 270: 8536-8539.

Brenowitz, S. and Castenholz, R. W. 1997. Long-term effects of UV and visible irradiance on natural populations of scytonemin-containing cyanobacterium (*Calothrix* sp.). *FEMS Microbiology Ecology* 24: 343-352.

Britt, A. B. 1995. Repair of DNA damage induced by ultraviolet radiation. *Plant Physiology* 108: 891-896.

Buma, A. G. J., Van Hannen, E. J., Roza, L., Veldhuis, M. J. W. and Gieskes, W. W. C. 1995. Monitoring ultraviolet-B induced DNA damage in individual diatom cells by immunofluorescent thymine dimer detection. *Jounal of Phycology* 51: 314-321.

Cadet, J., Sage, E. and Douki T. 2005. Ultraviolet radiation-mediated damage to cellular DNA. *Mutation Research* 571: 3-17.

Callone, A. I., Carignan, M., Montoya, N. G. and Carreto, J. I. 2006. Biotransformation of mycosporine like amino acids (MAAs) in the toxic dinoflagellate *Alaxandrium tamarense. Journal of Photochemistry and Photobiology B: Biology* 84: 204-212.

Campbell, D., Eriksson, M. -J., Öquist, G., Gustafsson, P. and Clarke, A. K. 1998. The cyanobacterium *Synechococcus* resists UV-B by exchanging photosystem II reaction-centre D1 proteins. Proceedings of the National Academy of Sciences (USA) 95: 364-369.

Carreto, J. I., Carignan, M. O., Daleo, G. and de Marco, S. G. 1990. Occurrence of mycosporine-like amino acids in the red tide dinoflagellate *Alexandrium excavatum*: UV-protective compounds? *Journal of Plankton Research* 12: 909-921.

Carell, T. and Epple, R. 1998. Repair of UV light induced DNA lesions: a comparative study with model compounds. *European Journal of Organic Chemistry* 1998: 1245-1258.

Castenholz, R. W. 1997. Multiple strategies for UV tolerance in cyanobacteria. *Spectrum* 10: 10-16.

Castenholz, R. W. and Garcia-Pichel, F. 2000. Cyanobacterial responses to UV-radiation. In: Whitton, B. A. and Potts, M. (eds.) *Ecology of Cyanobacteria: Their Diversity in Time and Space.* Kluwer Academic, Dordrecht, pp. 591-611.

Cockell, C. S. and Knowland, J. 1999. Ultraviolet radiation screening compounds. *Biological Reviews* 74: 311-345.

Cockell, C. S. and Rothschild, L. J. 1999. The effects of UV radiation A and B on diurnal variation in photosynthesis in three taxonomically and ecologically diverse microbial mats. *Photochemistry and Photobiology* 69: 203-210.

Conde, F. R., Churio, M. S. and Previtali, C. M. 2007. Experimental study of the excited- state properties and photostability of the mycosporine-like amino acid palythine in aqueous solution. *Photochemical Photobiological Sciences* 6: 669-674.

Crutzen, P. J. 1992. Ultraviolet on the increase. Nature 356: 104-105.

Daniel S., Cornelia S. and Fred Z. 2004. UV-A sunscreen from red algae for protection against premature skin aging. Cosmetic and Toiletries Manufacture Worldwide 139-143.

de Chazal, N. M. and Smith, G. D. 1994. Characterization of a brown *Nostoc* sp. from Java that is resistant to high light intensity and UV. *Microbiology* 140: 3183-3189.

Donkor, V. A., Damian, H. A. K. and Häder, D. -P. (1993) Effects of tropical solar radiation on the motility of filamentous cyanobacteria. *FEMS Microbiology Ecology* 12: 143-148.

Dunlap, W. C., Williams, D. M., Chalker, B. E. and Banaszak, A. T. 1989. Biochemical photoadaptations in vision: UV-absorbing pigments in fish eye tissues. *Comparative Biochemistry and Physiology* 93: 601-607.

Dunlap, W. C. and Shick, J. M. 1998. Ultraviolet radiation-absorbing mycosporine-like amino acids in coral reef organisms: a biochemical and environmental perspective. *Journal of Phycology* 34: 418-430.

Duniap, W. C. and Yamamoto, Y. 1995. Small-molecule antioxidants in marine organisms: antioxidant activity of mycosporine-glycine. *Comparative Biochemistry and Physiology* 112: 105-114.

Ehling-Schulz, M., Bilger, W. and Scherer, S. 1997. UV-B-induced synthesis of photoprotective pigments and extracellular polysaccharides in the terrestrial cyanobacterium *Nostoc commune*. *Journal of Bacteriology* 179: 1940-1945.

Ehling-Schulz, M., Schulz, S., Wait, R., Görg, A. and Scherer, S. 2002. The UV-B stimulon of the terrestrial cyanobacterium *Nostoc commune* comprises early shock proteins and late acclimation proteins. *Molecular Microbiology* 46: 827-843.

Eker, A. P., Kooiman, P., Hessels, J. K. C. and Yasui, A. 1990. DNA photoreactivating enzyme from the cyanobacterium *Anacystis nidulans*. *Journal of Biological Chemistry* 14: 8009-8015.

Essers, J., Vermeulen, W. and Houtsmuller A. B. 2006. DNA damage repair: anytime, anywhere? *Current Opinion in Cell Biology* 18: 240-246.

Fleming, E. D. and Castenholz, R. W. 2007. Effects of periodic desiccation on the synthesis of the UV-screening compound, scytonemin, in cyanobacteria. *Environmental Microbiology* 9: 1448-1455.

Frangeul, L., Quillardet, P., Castets, A. -M., Humbert, J. -F., Matthijs, H. C. P., Cortez, D., Tolonen, A., Zhang, C. -C., Gribaldo, S., Kehr, J. -C., Zilliges, Y., Ziemert, N., Becker, S., Talla, E., Latifi, A., Billault, A., Lepelletier, A., Dittmann, E., Bouchier, C., Tandeau de Marsac, N. 2008. Highly plastic genome of *Microcystis aeruginosa* PCC 7806, a ubiquitous toxic freshwater cyanobacterium. *BMC Genomics* 9: 274.

Friedberg, E. C., Walker, G. C. and Siede, W. 1995. DNA repair and mutagenesis. American Society of Microbiology Press, Washington, DC.

Gao, K., Yu, H., Brown, M. T. 2007. Solar PAR and UV radiation affects the physiology and morphology of the cyanobacterium *Anabaena* sp. PCC 7120. *Journal of Photochemistry and Photobiology B: Biology* 89: 117-124.

Gao, K., Li, P., Watanabe, T., Helbling, E. W. 2008. Combined effects of ultraviolet radiation and temperature on morphology, photosynthesis, and DNA of *Arthrospira* (*Spirulina platensis*) (Cyanophyta). *Journal of Phycology* 44: 777-786.

Garcia-Pichel, F. 1998. Solar ultraviolet light and the evolutionary history of cyanobacteria. Origins of Life and Evolution of the Biosphere 28: 321-347.

Garcia-Pichel, F. and Castenholz, R. W. 1991. Characterization and biological implications of scytonemin, a cyanobacterial sheath pigment. *Journal of Phycology* 27: 395-409.

Garcia-Pichel, F. and Castenholz, R. W. 1993. Occurrence of UV-absorbing mycosporine-like compounds among cyanobacterial isolates and an estimation of their screening capacity. *Applied and Environmental Microbiology* 59: 163-169.

Garcia-Pichel, F., Wingard, C. E. and Castenholz, R. W. 1993. Evidence regarding the UV sunscreen role of a mycosporine-like compound in the cyanobacterium *Gloeocapsa* sp. *Applied and Environmental Microbiology* 59: 170-176.

Gerber, S. and Häder, D. -P. 1995. Effects of enhanced solar irradiation on chlorophyll fluorescence and photosynthetic oxygen production of five species of phytoplankton. *FEMS Microbiology Ecology* 16: 32-42.

Gröniger, A., Sinha, R. P., Klisch, M. and Häder, D. -P. 2000. Photoprotective compounds in cyanobacteria, phytoplankton and macroalgae-a database. *Journal of Photochemistry and Photobiology B: Biology* 58: 115-122.

Häder, D. -P. and Worrest, R. C. 1991. Effects of enhanced solar radiation on aquatic ecosystems. *Photochemistry and Photobiology* 53: 717-725.

Häder, D. -P., Kumar, H. D., Smith, R. C. and Worrest, R. C. 2007. Effects of solar UV radiation on aquatic ecosystems and interactions with climate change. *Photochemical and Photobiological Sciences* 6: 267-285.

Han, T., Sinha, R. P., Häder, D. -P. 2003. Effects of intense PAR and UV radiation on photosynthesis, growth and pigmentation in the rice-field cyanobacterium *Anabaena* sp. *Photochemical and Photobiological Sciences* 2: 649-654.

Hargreaves, A., Taiwo, F. A., Duggan, O., Kirk, S. H., Ahmad, S. I. 2007. Near-ultraviolet photolysis of b-phenylpyruvic acid generates free radicals and results in DNA damage. Journal of Photochemistry and Photobiology B: Biology 89: 110-116.

He, Y. -Y., Klisch, M., Häder, D. -P. 2002. Adaptation of cyanobacteria to UV-B stress correlated with oxidative stress and oxidative damage. *Photochemistry and Photobiology* 76: 188-196.

He, Y. -Y. and Häder, D. -P. 2002a. UV-B-induced formation of reactive oxygen species and oxidative damage of the cyanobacterium *Anabaena* sp.: protective effects of ascorbic acid and N-acetyl-L-cysteine. *Journal of Photochemistry and Photobiology B: Biology* 66: 115-124.

He, Y. -Y. and Häder, D. -P. 2002b. Reactive oxygen species and UV-B: effect on cyanobacteria. *Photochemical and Photobiological Sciences* 1: 729-736.

He, Y. -Y., Häder, D. -P. 2002c. Involvement of reactive oxygen species in the UV-B damage to the cyanobacterium *Anabaena* sp. *Journal of Photochemistry and Photobiology B: Biology* 66: 73–80.

Helms, G. L., Moore, R. E., Niemczura, W. P., Patterson, G. M. L., Tomer, K. B. and Gross, M. L. 1988. Scytonemin A, a novel calcium antagonist from a blue-green alga. *Journal of Organic Chemistry* 53: 1298-1307.

Huang, L., McCluskey, M. P., Ni, H. and LaRossa, R. A. 2002. Global gene expression profiles of the cyanobacterium *Synechocystis* sp. strain PCC 6803 in response to irradiation with UV-B and white light. *Journal of Bacteriology* 184: 6845-6858.

Karentz, D., Cleaver, J. E. and Mitchell, D. L. 1991a. Cell survival characteristics and molecular responses of Antarctic phytoplankton to ultraviolet-B radiation. *Journal of Phycology* 27: 326-341.

Karentz, D., Mceuen, F. S., Land, M. C. and Dunlap, W. C. 1991b. Survey of mycosporine-like amino acid compounds in Antarctic marine organism: potential protection from ultraviolet exposure. *Marine Biology* 108: 157-166.

Kojima, K., Oshita, M., Nanjo, Y., Kasai, K., Tozawa, Y., Hayashi, H., Nishiyama, Y., 2007. Oxidation of elongation factor G inhibits the synthesis of the D1 protein of photosystem II. *Molecular Microbiology* 65: 936-947.

Kramlich, J.C., Linak, W.P., 1994. Nitrous oxide behaviour in the atmosphere, and in combustion and industrial systems. Progress in Energy and Combustion Science 20: 149-202.

Kumar, A., Tyagi, M. B., Srinivas, G., Singh, N., Kumar, H. D., Sinha, R. P. and Häder, D. -P. 1996. UV-B shielding role of $FeCl_3$ and certain cyanobacterial pigments. *Photochemistry and Photobiology* 64: 321-325.

Kumar, A., Tyagi, M. B., Jha, P. N., Srinivas, G. and Singh, A. 2003. Inactivation of cyanobacterial nitrogenase after exposure to ultraviolet-B radiation. *Current Microbiology* 46: 380-384.

Kumar, A., Tyagi, M. B., and Jha, P. N. 2004. Evidences showing ultraviolet-B radiation-induced damage of DNA in cyanobacteria and its detection by PCR assay. *Biochemical and Biophysical Research Communication* 318: 1025-1030.

Kumari, S., Rastogi, R. P., Singh, K. L., Singh, S. P., Sinha, R. P. 2008. DNA damage: detection strategies. *EXCLI Journal* 7: 44-62.

Lao, K. and Glazer, A. N. 1996. Ultraviolet-B photodestruction of a light-harvesting complex. Proceedings of the National Academy of Sciences USA 93: 5258-5263.

Lesser, M. P. 2008. Effects of ultraviolet radiation on productivity and nitrogen fixation in the cyanobacterium, *Anabaena* sp. (Newton's strain). *Hydrobiologia* 598: 1-9.

Levine, E. and Thiel, T. 1987. UV-inducible DNA repair in the cyanobacteria *Anabaena* sp. *Journal of Bacteriology* 169: 3988-3993.

Lindahl, T. 1993. Instability and decay of the primary structure of DNA. *Nature* 362: 709-715.

Mackerness, S. A. H., Jordan, B. R. and Thomas, B. 1999. Reactive oxygen species in the regulation of photosynthetic genes by ultraviolet radiation (UV-B: 280-320 nm) in green and etiolated buds of pea (*Pisum sativum* L.). *Journal of Photochemistry and Photobiology B: Biology* 48: 180-188.

Mate, Z., Sass, L., Szekeres, M., Vass, I. and Nagy, F. 1998. UV-B-induced differential transcription of *psbA* genes encoding the D1 protein of photosystem II in the cyanobacterium *Synechocystis* 6803. *Journal of Biological Chemistry* 273: 17439-17444.

McClintock, J. B. and Karentz, D. 1997. Mycosporine-like amino acids in 38 species of subtidal marine organisms from McMurdo Sound, Antarctica. *Antarctic Science* 9: 392-398.

Misonou T., Saitoh J., Oshiba S., Tokitomo Y., Maegawa M., Inoue Y., Hori H., and Sakurai T. 2003. UV-Absorbing substance in the red alga *Porphyra yezoensis* (Bangiales, Rhodophyta) block thymine photodimer production. *Marine Biotechnology* 5: 194-200.

Mitchell, D. L. and Karentz, D. 1993. The induction and repair of DNA photodamage in the environment. In: Young, A. R., Björn, L., Moan, J. and Nultsch, W. (eds.), *Environmental UV Photobiology*, pp. 345-377. Plenum Press, London.

Newton, J. W., Tyler, D. D. and Slodki, M. E. 1979. Effect of ultraviolet-B (280-320 nm) radiation on blue-green algae (cyanobacteria): possible biological indicators of stratospheric ozone depletion. *Applied and Environmental Microbiology* 37: 1137-1141.

Ning, S. B., Guo, H. L., Wang, L., Song, Y. C. 2002. Salt stress induces programmed cell death in prokaryotic organism *Anabaena*. *Journal of Applied Microbiology* 93: 15-28.

Niyogi, K. K. 1999. Photoprotection revisited: genetics and molecular approaches. Annual *Review of Plant Physiology* 50: 333-359.

Oren, A. 1997. Mycosporine-like amino acids as osmotic solutes in a community of halophilic cyanobacteria. *Geomicrobiology Journal* 14: 231-240.

Oren, A. and Gunde-Cimerman, N. 2007. Mycosporines and mycosporine-like amino acids: UV protectants or multipurpose secondary metabolites? *FEMS Microbiology* Letters 269: 1-10.

Oyamada, C., Kaneniwa, M., Ebitani, K., Murata, M., Ishihara, K. 2008. Mycosporine-like amino acids extracted from scallop (*Patinopecten yessoensis*) ovaries: UV protection and growth stimulation activities on human cells. *Marine Biotechnology* 10: 141-150.

Phoenix, V. R., Bennett, P. C., Engel, A. S., Tyler, S. W., Ferris, F. G., 2006. Chilean high altitude hot-spring sinters: a model system for UV screening mechanisms by early Precambrian cyanobacteria. *Geobiology* 4: 15-28.

Pierson, B. K, Mitchell, H. K. and Ruff-Roberts, A. L. 1993. *Chloroflexus aurantiacus* and ultraviolet radiation: implications for archean shallow water stromatolites. *Origins of Life and Evolution of the Biosphere* 23: 243-260.

Portwich, A. and Garcia-Pichel, F. 1999. Ultraviolet and osmotic stresses induce and regulate the synthesis of mycosporines in the cyanobacterium *Chlorogloeopsis* PCC 6912. *Archives of Microbiology* 172: 187-192.

Portwich, A. and Garcia-Pichel, F. 2000. A novel prokaryotic UV-B photoreceptor in the cyanobacterium *Chlorogloeopsis* PCC 6912. *Photochemistry and Photobiology* 71: 493-498.

Prasad, S. M. and Zeeshan, M. 2004. Effect of UV-B and monocrotophos, singly and in combination, on photosynthetic activity and growth of non-heterocystous cyanobacterium *Plectonema boryanum*. *Environmental and Experimental Botany* 52: 175-184.

Proteau, P. J., Gerwick, W. H., Garcia-Pichel, F. and Castenholz, R. W. 1993. The structure of scytonemin, an ultraviolet sunscreen pigment from the sheath of cyanobacteria, *Experimentia* 49: 825-829.

Quesada, A. and Vincent, W. F. 1997. Strategies of adaptation by Antarctic cyanobacteria to ultraviolet radiation. *European Journal of Phycology* 32: 335-342.

Quesada, A., Vincent, W. F. and Lean-David, R. S. 1999. Community and pigment structure of arctic cyanobacterial assemblages: the occurrence and distribution of UV-absorbing compounds. *FEMS Microbiology Ecology* 28: 315-323.

Ravanat, J. L., Douki, T. and Cadet, J. 2001. Direct and indirect effects of UV radiation on DNA and its components. *Journal of Photochemistry and Photobiology B: Biology* 63: 88-102.

Richter, P., Krywult, M., Sinha, R. P. and Häder, D. -P. 1999. Calcium signals from heterocysts of *Anabaena* sp. after UV irradiation. *Journal of Plant Physiology* 154: 137-139.

Richter, P. R., Sinha, R. P. and Häder, D. -P. 2006. Scytonemin-rich epilithic cyanobacteria survive acetone treatment. *Current Trends in Microbiology* 2: 13-19.

Ross, C., Santiago-Vázquez, L., Paul, V. 2006. Toxin release in response to oxidative stress and programmed cell death in the cyanobacterium *Microcystis aeruginosa. Aquatic Toxicology* 78: 66-73.

Sass, L., Spetea, C., Mate, Z., Nagy, F. and Vass, I. 1997. Repair of UV-B induced damage of photosystem II via *de novo* synthesis of the D1 and D2 reaction centre subunits in *Synechocystis* sp. PCC6803. *Photosynthetic Research* 54: 55-62.

Scherer, S., Chen, T. W. and Böger, P. 1988. A new UV-A/B protecting pigment in the terrestrial cyanobacterium *Nostoc commune. Plant Physiology* 88: 1055-1057.

Shibata, H., Baba, K. and Ochiai, H. 1991. Near UV irradiation induces shock proteins in *Anacystis nidulans* R-2; possible role of oxygen. *Plant Cell Physiology* 32: 771-776.

Singh, S. P., Kumari, S., Rastogi, R. P., Singh, K. L., Sinha, R.P. 2008a. Mycosporine-like amino acids (MAAs): chemical structure, biosynthesis and significance as UV-absorbing/ screening compounds. *Indian Journal of Experimental Biology* 46: 7-17.

Singh, S. P., Klisch, M., Sinha, R. P., Häder, D. -P. 2008b. Effects of abiotic stressors on synthesis of the mycosporine-like amino acid shinorine in the cyanobacterium *Anabaena variabilis* PCC 7937. *Photochemistry and Photobiology* 84: 1500-1505.

Singh, S. P., Klisch, M., Häder, D. P., Sinha, R. P. 2008c. Role of various growth media on shinorine (mycosporine-like amino acid) concentration and photosynthetic yield in *Anabaena variabilis* PCC 7937. *World Journal of Microbiology and Biotechnology* 24: 3111-3115.

Singh, S. P., Sinha, R. P., Klisch, M., Häder, D. P. 2008d. Mycosporine-like amino acids (MAAs) profile of a rice-field cyanobacterium *Anabaena doliolum* as influenced by PAR and UVR. Planta 229: 225-233.

Sinha, R. P., Kumar, H. D., Kumar, A. and Häder, D. -P. 1995. Effects of UV-B irradiation on growth, survival, pigmentation and nitrogen metabolism enzymes in cyanobacteria. Acta Protozoologica 34: 187-192.

Sinha, R. P., Singh, N., Kumar, A., Kumar, H. D., Häder, M. and Häder, D. P. 1996. Effect of UV radiation on certain physiological and biochemical processes in cyanobacteria. *Journal of Photochemistry and Photobiology B: Biology* 32: 107-113.

Sinha, R. P. and Häder, D. P. 1998. Effects of ultraviolet-B radiation in three rice field cyanobacteria. *Journal of Plant Physiology* 153: 763-769.

Sinha, R. P., Klisch, M., Vaishampayan, A. and Häder, D. P. 1999a. Biochemical and spectroscopic characterization of the cyanobacterium *Lyngbya* sp. inhabiting mango (*Mangifera indica*) trees: presence of an ultraviolet-absorbing pigment, scytonemin. Acta Protozoologica 38: 291-298.

Sinha, R. P., Klisch, M. and Häder, D. P. 1999b. Induction of a mycosporine-like amino acid (MAA) in the rice-field cyanobacterium *Anabaena* sp. by UV irradiation. *Journal of Photochemistry and Photobiology B: Biology* 52: 59-64.

Sinha, R. P. and Häder, D. -P. 2001. Ultraviolet-B radiation effects on cyanobacteria and the role of photoprotective compounds in mitigating UV-B toxicity. *Trends in Photochemistry and Photobiology* 8: 117-127.

Sinha, R. P., Dautz, M. and Häder, D. -P. 2001a. A simple and efficient method for the quantitative analysis of thymine dimers in cyanobacteria, phytoplankton and macroalgae. *Acta Protozoologica* 40: 187-195.

Sinha, R. P., Klisch, M., Helbling, E. W. and Häder, D. -P. 2001b. Induction of mycosporine-like amino acids (MAAs) in cyanobacteria by solar ultraviolet-B radiation. *Journal of Photochemistry and Photobiology* B: Biology 60: 129-135.

Sinha, R. P. and Häder, D. P. 2002a. Life under solar UV radiation in aquatic organisms. *Advances in Space Research* 30: 1547-1556.

Sinha, R. P. and Häder, D. P. 2002b. UV-induced DNA damage and repair: a review. *Photochemical and Photobiological Sciences* 1: 225-236.

Sinha, R. P. and Häder, D. P. 2006. Impact of UV radiation on rice field cyanobacteria: role of photoprotective compounds, In: F. Ghetti *et al.* (eds.) *Environmental UV Radiation: Impact on Ecosystems and Human Health,* Springer, The Netherlands, pp. 217-230.

Sinha, R. P., Singh, S. P. and Häder, D. -P. 2007. Database on mycosporines and mycosporine-like amino acids (MAAs) in fungi, cyanobacteria, macroalgae, phytoplankton and animals. *Journal of Photochemistry and Photobiology B: Biology* 89: 29-35.

Sinha, R. P. and Häder, D. P. 2008. UV-protectants in cyanobacteria. Plant Science 174: 278-789.

Smith, R. C., Prezelin, B. B., Baker, K. S., Bidigare, R. R., Boucher, N. P., Coley, T., Karentz, D., MacIntyre, S., Matlick, H.A., Menzies, D., Ondrusek, M., Wan, Z. and Waters, K. J. 1992. Ozone depletion: ultraviolet radiation and phytoplankton biology in Antarctic waters. *Science* 255: 952-959.

Soule, T., Stout, V., Swingley, W. D., Meeks, J. C., Garcia-Pichel, F. 2007. Molecular genetics and genomic analysis of scytonemin biosynthesis in *Nostoc punctiforme* ATCC 29133. *Journal of Bacteriology* 189: 4465-4472.

Starcevic, A., Akthar, S., Dunlap, W. C., Shick, J. M., Hranueli, D., Cullum, J., Long, P. F. 2008. Enzymes of the shikimic acid pathway encoded in the genome of a basal metazoan, *Nematostella vectensis*, have microbial origins. Proceedings of the National Academy of Sciences USA 105: 2533-2537.

Stevenson, C. S., Capper, E. A., Roshak, A. K., Marquez, B., Eichman, C., Jackson, J. R., Mattern, M., Gerwick, W. H., Jacobs, R. S. and Marshall, L. A. 2002a. The identification and characterization of the marine natural product scytonemin as a novel antiproliferative pharmacophore. The *Journal of Pharmacology and Experimental Therapeutics* 303: 858-866.

Stevenson, C. S., Capper, E. A., Roshak, A. K., Marquez, B., Grace, K., Gerwick, W. H., Jacobs, R. S. and Marshall, L. A. 2002b. Scytonemin- a marine natural product inhibitor of kinases key in hyperproliferative inflammatory diseases. *Inflammation Research* 51: 112-114.

Suh, H. J., Lee, H. W. and Jung, J. 2003. Mycosporine-glycine protects biological systems against photodynamic damage by quenching singlet oxygen with a high efficiency, *Photochemistry and Photobiology* 78: 109-113.

Tel-Or, E., Huflejt, M. E., Packer, L. 1986. Hydroperoxide metabolism in cyanobacteria. *Archives in Biochemistry and Biophysics* 246: 396-402.

Torres, A., Enk, C. D., Hochberg, M. and Srebnik, M. 2006. Porphyra-334, a potential natural source for UVA protective sunscreens. *Photochemical and Photobiological Sciences* 5: 432-443.

Tyagi, R., Srinivas, G., Vyas, D., Kumar, A. and Kumar, H. D. 1992. Differential effect of ultraviolet-B radiation on certain metabolic processes in a chromatically adapting *Nostoc. Photochemistry and Photobiology* 55: 401-407.

Wang, Y., and Tang, X. 2008. Interactions between *Prorocentrum donghaiense* Lu and *Scrippsiella trochoidea* (Stein) Loeblich III under laboratory culture. *Harmful Algae* 7: 65-75.

Wang, G., Chen, K., Chen, L., Hu, C., Zhang, D. and Liu, Y. 2008. The involvement of the antioxidant system in protection of desert cyanobacterium *Nostoc* sp. against UV-B radiation and the effects of exogenous antioxidants. *Ecotoxicology and Environmental Safety* 69: 150-157.

Wilson, M. I., Ghosh, S., Gerhardt, K. E., Holland, N., Sudhakar Babu, T., Edeeman, M., Dumbroff, E. B. and Greenberg, B. M. 1995. *In vivo* photomodification of ribulose-1,5-bisphosphate carboxylase/ oxygenase holoenzyme by ultraviolet-B radiation. *Plant Physiology* 109: 221-229.

Wu, H., Gao, K., Villafañe, V. E., Watanabe, T. and Helbling, E. W. 2005. Effects of solar UV radiation on morphology and photosynthesis of filamentous cyanobacterium *Arthrospira platensis*. *Applied and Environmental Microbiology* 71: 5004–5013.

Yoshiki, M., Tsuge, K., Tsuruta, Y., Yoshimura, T., Koganemaru, K., Sumi, T., Matsui, T. and Matsumoto, K. 2009. Production of new antioxidant compound from mycosporine-like amino acid, porphyra-334 by heat treatment. *Food Chemistry* 113: 1127-1132.

Yuan, Y. V., Westcott, N. D., Huc, C. and Kitts, D. D. 2009. Mycosporine-like amino acid composition of the edible red alga, *Palmaria palmata* (dulse) harvested from the west and east coasts of Grand Manan Island, New Brunswick. *Food Chemistry* 112: 321–328.

Zhang, L., Li, L. and Wu, Q. 2007. Protective effects of mycosporine-like amino acids of *Synechocystis* sp. PCC 6803 and their partial characterization. *Journal of Photochemistry and Photobiology B: Biology* 86: 240-245.

Zhao, W., Ye, Z., Zhao, J., 2007. RbrA, a cyanobacterial rubrerythrin, functions as a FNR-dependent peroxidase in heterocysts in protection of nitrogenase from damage by hydrogen peroxide in *Anabaena* sp. PCC 7120. *Molecular Microbiology* 66: 1219-1230.

Zhou, W., Tang, X., Xiao, H., Wang, Y. and Wang, R. 2009. Response of marine microalgae, heterotrophic bacteria and their relationship to enhanced UV-radiation. *Journal of Ocean University of China* 8: 35-38.

Chapter 3

Seed Dormancy—An Ecological Adaptation

M.P. Trivedi, Rachna Kumari and U.K. Sinha

Seeds occupy a critical position in the life history of higher plants. The success with which the new individual established is largely determined by the time, the place and the vigour of the young seedlings and physiological and biological features of the seeds. The success at establishment also depends on seed responses to the environment and its biochemical level. Its food reserves are able to sustain the young plant in the early stages of growth before it becomes a self-sustaining independent entity.

There are seeds which donot germinate even if the various factors influencing seed germination are favourable. Such seeds whose germination is inhibited due to some internal limitations are known as dormant. For such seeds to germinate, they must be hydrated in conditions which encourage metabolism such as suitable temperature and presence of oxygen.

Seed dormancy can be easily visualised in wild legumes like *Alysicarpus spp., Melilotus alba, M. indica, Indigofera astragalina, I. glandulosa, I. linnaei, I. linifolia, Medicago lupulina, M. denticulata, Clitoria ternatea,* Apocynaceous member *Rauvolfia serpentina* and Caesalpiniaceous member *Cassia obtussifolia* and *Cassia tora*.

Dormancy is also called rest. Imposed dormancy is also known as quiescence. It is the inability of a viable seed to germinate because the environmental factors required for it are not available.

PRIMARY AND SECONDARY DORMANCY

Primary dormancy is due to some intrinsic factors related to the seeds as against secondary dormancy which develops due to some changes occuring during storage such as in *Taxus* and *Fraxinus*. These seeds may germinate immediately after they are shed off

Department of Botany, Patna Science College, Patna University, Patna - 800 005 and environmental conditions are favourable. eg *Rauvolfia tetraphylla*. They donot germinate if they are stored.

CAUSES OF SEED DORMANCY

Hard and Thick Seedcoat

It inhibits entry of water and oxygen. Thus, the seeds are not hydrated. Non availability of oxygen and increased concentration of Co_2 in the interior of seed repress respiration and hence germination.

Seeds of many wild leguminous plants are hard, resistant to abrasion and covered with a wax like layer. Sometimes we find suberised strophioler cleft (counter palisade tissue). Hilum and chalaza are also blocked by special cells.

Besides the usual coat dormancy, an innate dormancy caused by inhibitor was evident particularly in *Alysicarpus bupleurifolius*. Five months old seeds did not show innate dormancy and also were with diminished coat dormancy (Pandey & Goel, 2003). Ram Krishnana and Khosla 1971 worked out seed dormancy in *Digitaria adscendens* and *Echinochloa colona* with special reference to concerning structure.

Immature Embryo

Sometimes there is immature embryo after shedding of the seeds. Thus, there is no question of germination without complete development of embryo i.e. *Ginkgo biloba, Anemone neumorosa, Caltha palustris*.

The members of Orchidaceae have similar features. In *Fraxinus excelsior*, the embryo is fully developed at the time of seed shedding but requires considerable growth before it is able to form a seedling. The time required for maturity of embryo varies from few days to several months. During this after ripening period, several morphological, anatomical, and biochemical changes take place.

Germination Inhibitors

Such inhibitors are parascorbic acid and coumerin (an unsaturated lactone). They are present in fruit pulp or seedcoat. In a number of cereals the dormancy is due to glumes. The hull of *Avena fatua* does not allow the leaching of inhibitors present in caryopsis. In *Momordica charantia* the germination inhibitors are present in fruit pulp. There are other germination inhibitors such as cyanide, dinitrophenol, azide, chloride, hydroxylamine, flouride etc.

Cycocel, Phosphon-D, AMO-1618 and phenolic compounds are growth retardants and inhibit germination.

Abscisic acid is also a germination inhibitor and its content and dormancy can be correlated in *Fraxinus americana*.

Temperature Requirements

Optimum temperature is the prime requirement for seed germination. Some seeds require either high or low temperature before germination. *Polygonum* spp and some conifers require low temperature in presence of oxygen in moist condition for weeks to months for germination. Sometimes alternating high and low temperatures are required.

Specific Requirement of Light

Some seeds require specific light for germination e.g. - *Lactuca sativa* which germinates in red light in comparison to seeds of *Nigella sativa*, which germinates in dark.

High Osmotic Concentration

The seeds of *Atriplex* have a very high osmotic concentration. This prevents their germination. Such seeds germinate only when some solutes come out from seeds due to optimum rainfall.

Methods of Breaking Seed Dormancy

There are various treatments which can break the dormancy of seeds. Such treatments are as below :

1. Mechanical Scarification

Scarification is breaking the barrier of seedcoat. It is done to weaken the seedcoat with the help of abrasive papers, sand, needle or blade. Shaking of seeds is also helpful if strophioler plug is subersied. eg *Crotalaria.*

In *Rauvolfia serpentina* seeds are mechanically scarified with blade and only half coats are removed. Care is taken to protect the embryonal portion. (Sinha *et al*, 2008). Partial cut of seedcoat by blade also removes dormancy in *Clitoria ternatea*.

2. Chemical Scarification

It is done by Conc H_2SO_4, HCl, HNO_3, alcohol, thiourea and ascorbic acid. Sulphuric acid has been observed to be most successful and is widely used as a dormancy breaking chemical.

Seeds of different legumes namely *Melilotus alba* Lamk, *M. indica*. All, *Indigofera linnaei* Ali, *I. linifolia* (Linn.) Retz, *Medicago lupulina L.* and *M. denticulata* Willd, *Indigofera astragalina* D.C. Prodr. and *I. glandulosa* Willd, all have seedcoat dormancy. Kumari 1993, worked out dormancy in these wild legumes.

In control *I. linnaei* showed 18%, *I. linifolea* 3%, *I. astragalena, Melilotus alba* and *Medicago denticulata*, 8% *M. Indica* and *M.lupulina* 10% germination. They were scarified with conc. H_2SO_4 for 5, to 40 minutes and were washed thoroughly in running water for 2h to remove traces of H_2SO_4. Seeds, thus, washed were kept for germination. Seeds treated with conc. H_2SO_4 for 20 minutes germinated upto 91% in *M. indica*, 100% in *M. alba* and *M. lupulina* and 88% in *I. linnaei. I. linifolia* showed 88% in seeds treated for 10 minutes only. Seeds treated for 30 minutes showed 100% germination in *I.astragalina* and *I. glandulosa. M. denticulata* showed maximum germination (50%) after treated for 30 minutes. It is interesting that seeds even tolerate 30-45 minutes of treatment with conc. H_2SO_4 although their germination potential declined (Tables 1&2).

On perusal of results, it appears that *M. Indica, I. linnaei* and *M. denticulata* have an extremely hard seedcoat followed by *M. lupulina, M. alba, I. linifolia, I. astragalina and I. glandulosa.* H_2SO_4 reduces the content of inhibitors. Similar effect of enhanced germinatibility has been obtained by Agarwal & Vyas (1970); Yadav *et al* (1979). Germination in *Melilotus, Medicago* and *Indigofera* spp. after scarification (Tables 1 & 2) to the extent of 50-100% amply proves that embryo is fully ripe and able to germinate if favourable conditions are provided to them. It also proves that dormancy is due to hard seedcoat which prevents the germination of seeds during summer.

TABLE 1

Effect of scarification (conc. H_2SO_4) on germination percentage

Duration of treatment (Min)	Germination percentage			
	I. linnaei	*I. linifolia*	*I. astragalina*	*I. glandulosa*
05	26.0	40.0	20.0	25.0
10	40.0	88.0	50.0	50.0
15	72.0	80.0	85.0	85.5
20	88.0	50.0	90.6	87.76
25	44.0	08.0	95.0	90.0
30	30.0	00	100.0	100.0
35	25.0	00	60.0	60.0
40	20.0	00	45.0	40.0

TABLE 2

Effect of scarification (conc. H_2SO_4) on germination percentage

Duration of treatment (Min)	Germination percentage			
	M. alba	*M. Indica*	*M. denticulata*	*M. lupulina*
05	37.6	50.6	20.0	50.0
10	70.3	60.3	25.0	60.0
15	75.0	86.6	30.0	100.0
20	100.0	90.6	30.6	100.0
25	90.0	85.5	40.4	100.0
30	86.6	70.0	50.0	100.0
35	60.0	50.0	45.0	100.0
40	40.0	30.0	10.0	80.0

In nature scarification is achieved by microbial action (Wareing, 1963) and passage of seed through digestive tract of bird or other animals or movement by water across sand or rock.

Alcohol treatment is affective for seeds of family Caesal piniaceae. Dormancy in cotton seed and *Prosopis juliflora* can be removed by conc. H_2SO_4. Chaudhary & Sinha (1989) worked out effect of thiourea, ascorbic acid and their interaction on seed germination of *Portulaca quadrifida* Linn. They reported that thiourea possesses the ability of breaking dormancy and serves as germination promoter in light whereas ascorbic acid has been reported to play dual role — (a) Synergism with thiourea in stimulating germination and (b) stimulation of germination in dark with increasing concentration.

Organic solvent like xylene, acetone and alcohol also caused breakage of dormancy to certain extent in *Alysicarpus bupleurifolius, A. monilifer* and *A. vaginalis.*

STRATIFICATION

Effect of low temeperature (0-5^{0}C) for a period ranging from a couple of weeks to a few months on moistened seed removes dormancy in peach, plum, cherry and apricot. Low temperature induces anatomical and biochemical changes like increased gibberellin content and mobilization of food materials towards embryo.

Pressure

It is subjecting of seeds to a high hydraulic pressure (about 2000 atm) at a temperature of 18-20^{0}C for 5-20 minutes. (Bhatia and Parasar, 1991). The applied pressure increases permeability and weakens the seedcoat. This method is suitable for *Melilotus alba* and *Medicago sativa.*

Exposure to Alternate Temperatures

Dormancy can be overcome by subjecting the seeds to alternately high and low temperatures. This requirement is available in nature in the form of diurnal change. Alternate temperature can change the permeability of seeds to gases and cause phytochrome conversion and its recycling. (Bhatia and Parasar, 1991).

After Ripening Storage

Seeds obtained after plant harvests are sealed in gunny bags or in earthen pot. They are suitably placed for required months. Co_2 released in storage enhances temperature that allows embryo to develop. After 6 months to one year seeds are taken out for germination. Higher temperature of storage (25^{0}C and room temperature) was helpful in removal of dormancy upto an extent of 30% in *A. bupleurifolius.*

GA_3 and Cytokinin Treatment

Gibberellins, 2-chloroethanol or ethylene chlorohydrine, Thiourea and Cytokinin are capable to break dormancy in specific cases. GA_3 induces the synthesis of a-amylase in cereals which digests the starch present in endosperm and make it available to embryo. GA_3 also substitutes low temperature requirement. Ethylene also promotes germination of seeds in *Striga asiaticus* (Srivastava, 2005).

Unscarified seeds of *I.linifolia* displayed 40% germination at 10 ppm of GA_3 and Cytokinin. In *I. astragalina*, unscarified seeds showed maximum germination (45%) at 50 ppm of GA_3. Unscarified seeds of *I. glandulosa* showed 25% germination at 10 ppm of GA_3. (Table 3). Mechanically scarified seeds of *Rauvolfia serpentina* germinate upto 54%. The scarified seeds after treatment with 100 ppm of GA_3 and Cytokinin showed enhanced germination (Table-4).

The tables 3 & 4 show lower concentration of GA_3 and moderate concentration of Cytokinin as suitable for breaking seed dormancy. MH was considered to be the best to increase the seed germination percentage of *Cassia obtussifolia* (Singh & Murty, 1987).

TABLE 3

Effect of GA_3 and cytokinin on germination percentage

	Concentration (ppm)	*I. astragalina* Unsc.	*I. astragalina* SC	*I. glandulosa* Unsc.	*I. glandulosa* SC	*M. alba* Unsc.	*M. alba* SC	*M. indica* Unsc.	*M. indica* SC
	10	10	10	25	100	10	25	65	12
	25	25	17	17	100	5	22	50	16
GA_3	50	45	7	7	90	-	16	50	10
	75	10	5	5	80	-	10	20	10
	100	5	2	2	60	-	10	15	10
Cy	10	8	90	10	100	10	25	10	93
to	25	8	60	-	75	40	55	35	85
ki	50	5	50	-	65	30	20	5	65
ni	75	-	40	-	40	5	8	5	60
n	100	-	10	-	50	5	5	5	50

Unsc = Unscarified, Sc = Scarified

TABLE 4

Effect of growth regulators on germination potential of *Rauvolfia serpentina*

Concentration (ppm)	Initiation period (days)	Rate of germination per day in Sc. Seeds (%) Gibberellic Acid	Rate of germination per day in Sc. Seeds (%) Cytokinin	Percentage of germination in Sc. seeds Gibberellic Acid	Percentage of germination in Sc. seeds Cytokinin
Control	1	39.2	39.2	54.0	54.0
10	1	54.2	72.5	73.1	90.4
25	1	42.5	76.8	51.6	95.2
50	1	45.2	71.9	60.2	90.4
100	1	54.8	82.7	77.4	95.2

Sc.= Scarified

REMOVAL OF GERMINATION INHIBITOR

This is possible due to repeated washing with water and treatment of seeds at lower temperature. In *Alysicarpus* spp. freshly harvested seeds failed to germinate. They were placed in running water for 48h, that, water soluble inhibitor, if any, could be washed out. On leaching, 30, 7.5 and 2.15% of seeds of *Alysicarpus bupleurifolius, A. monilifer* and *A. vaginalis* respectively germinated (Pandey & Goel, 2003).

SEED DORMANCY - AN ECOLOGICAL ADAPTATION

It is an ecological adaptation and growth ceasation which often confer upon some species a selective advantage in distribution.

Leguminous seeds germinate after the rainy season is over and flower and fruit during February and March. During rest months of the year they remain dormant. Dormancy allows seeds to pass through unfavourable conditions like drought and cold without any injury. It is a factor affecting seed survival in soil and is an important component of seed dispersal in many plants. The germination inhibitors in seedcoat of desert plants allow them to remain inactive during the dry periods. When inhibitors are leached out due to rainfall, the seeds germinate. Dormancy also protects the species from extinction in the year of drought. Some weeds of this type are very difficult to be eradicated because new plants will continue to grow over the years. Delayed germination of grains and seeds can help us in their storage and for their proper distribution, both in time and space.

Various synthetic chemicals have been developed which will cause prolongation of the period of rest or dormancy. i.e. 2, 4-D, NAA, Maleic hydrazide, 2, 4, 5-T etc.

REFERENCES

Agarwal, S.K. 1971, Effect of thiourea and ascorbic acid on seed germination. *J. Indian Bot. Soc.* 50 : 374-376.

Agarwal, S.K. and Vyas, L.N. 1970. Studies in the extent and role of dormancy in the seeds of *Indigofera astragalina* DC Prodr. *J. Indian Bot. Soc.* 49 (1-4) : 158-163.

Bhatia, K.N. and Parasar, A.N. 1991. Plant physiology a modern treatise, True Man Book Company, Jalandhar City.

Chaudhary, S.K. and Sinha, R.P. 1989. Effect of thiourea, ascorbic acid and their interaction on seed germination of *Portulaca quadrifida* Linn., *Biojournal* 1 (1) : 57-62.

Kumari, 1993. Effect of growth regulators on growth and nodulation of some fabaceous weeds, Ph.D. thesis, Patna University, Patna.

Pandey, B.N. and Goel R.K. 2003. Studies on dormancy and germination of seeds of some species of *Alysicarpus* DC. – The Fabaceous weeds of Semi-arid grasslands. In environmentaal Pollution : Problems and Management, Jaspal Prakashan, Patna 152-162.

Ramakrishnan, P.S. and Khosla, A.K. 1971. Seed dormancy in *Digitaria adscendens* and *Echinochloa colona* with special reference to covering structure. *Trop Ecol* 12 (1) : 112-122.

Singh Chhavi and Murty Y.S. 1987. Effect of some growth regulators on the seed germination and seedling growth of *Cassia obtusifolia* Linn., *Acta Botanica Indica* 15 : 74-79.

Sinha, U.K., Trivedi M.P. and Kumari Rachna, 2008. Seed germination studies in *Rauvolfia serpentina, Manthan.* 1 : 39-43

Srivastava, H.P. 2005. Plant physiology, Rastogi publication Meerut.

Wareing, P.F. 1963. The germination of seeds. In Vistas in Botany vol. III (Turriled) Pergamon Press, pp. 195.

Yadav, R.D.R., Verma, D.P.S., Singh, Amar and Sastri, J.A. 1979. Germination studies on *Medicago* spp. following seed treatment. *Seed Research* 7 (1) : 71-76.

❒❒❒

Chapter 4

Water Water Every Where, But not a Drop to Drink

ANIL K. DWIVEDI*, UDAY BHAN PRAJAPATI* AND SHASHI**

INTRODUCTION

Nature has provided the variety of living and non-living component on Earth. Earth is a special planet, which has air, water, soil, minerals, ores, grassland, river, ocean, desert, mountain, forest, flora etc. All natural resources are the precious gift of nature. This has been possible only because of appropriate distance between Earth and Sun and also due to its strong gravitational force, which keeps the atmospheric air, bound to it. Next to air, water is the most essential component of our ecosystem.

To sustain life on Earth in all its totality, water should be carefully managed in its natural habitat. This is accomplished by managing life on Earth, as life is both a consequence and cause of water on land. It is universally known that water is the basis of life on Earth. Our very survival on Earth essentially depends on three basic resources water, soil, and air, nature's three valuable gifts to the mankind. The abundance of water on Earth is a unique feature that distinguishes our "blue planet" from other in the solar system. Approximately 70.8 per cent (97 per cent of it being sea water and 3 per cent fresh water) of the Earth is covered by water and only 29.2 per cent is landmass. Earth's solar orbit, volcanism, gravity, greenhouse effect, magnetic field and oxygen rich atmosphere seem to combine to make Earth a water planet.

A hydrosphere (Greek hydro means water) in physical geography described the collective mass of water found on, under and over the surface of a planet. The Earth's hydrosphere consists of water in its liquid, solid and gaseous form. It is unique liquid in the sense that without it, no life is possible. Though on the global scale, there is

* Department of Botany, Arya Mahila (P.G.) College, Shahjahanpur - 242 001, U.P.

** Plant Resources Research Foundation, 72-Faridi Colony, Shahjahanpur - 242001, U.P.

abundance of water but the main problem is the availability of water in the right place at the right time and in the right form. The concern on water scarcity, at least among the 'wise' amongst us, is on the increase, despite the fact that water is renewable, maintained by the solar energy driven hydrological cycle. Two expression of concern are worth quoting here. (Ed. Issue. *Current Sc.* 2005).

"Born in a water-rich environment we have never really learnt how important water is to us.........We have spent it with shameful and unbecoming hast.........Everywhere we have poured filth into it."

-William Anshworth

"Our entire society rest upon and is depend upon –our water, our land, our forest and our mineral resources. How we use these resources influence our health, security, economic and well being"

-John F. Kennedy

In contemporary society water is a major theme of scientific, economic, political, social and human debate. There is widespread concern that the future of peace and prosperity in a finite Earth will depend on our civilized ability to share the limited water resources of the world.

Water is found in large storage in all the three forms like solid in the form of ice on the North and South poles and on the top of high altitude mountain, liquid in saline water, in ocean and freshwater in lakes, river, pools, stream, dams, wetland and in soil pores as well as subsoil water table and deep underground aquifers and in gaseous form in invisible vapour form like cloud and fog.

The necessity of water to life is so important that water has been accepted as a natural right. The United Nations Human Right (Yojna, 2007) statement on right to water says, "The human right to water entitles every one to sufficient, safe, acceptable, physiologically accessible and affordable water for personal and domestic uses. An adequate amount of safe water is necessary to prevent death from dehydration, reduce the risk of water related disease and provide for consumption, cooking, personal and domestic hygienic requirements."

The Union government declared 2007 as "water year" to focus on water related issues and launched a massive awareness program in the country. The 11th five-year plan wills priorities development of water resources.

Current understanding of the hydrological cycle show that, this phenomenon involving a fraction of one percent of all water on the Earth, sustain all life. The hydrological nutrient and erosion cycles on which all living communities depend are delicately interlinked. Undue human intervention with these cycles can force them towards new equilibrium that may seriously stress or otherwise endanger existing communities. Global worming, salinization of soil, desertification, endangerment of species, contamination of freshwater supplies and so on are examples. These impacts are affecting developed and developing nation alike.

PROPERTIES OF WATER

Life originated and passed through early evolution in water and continues to exist on Earth only because of unique physico-chemical properties of water, Remember the saying that:

"Earthly organisms are aqueous system on a watery planet"

At room temperature water is a liquid. Pure liquid water is colorless, transparent, testless, odourless and very diffusible due to small size of its molecules. Melting point of pure water is 0°C and boiling point 100°C.The freezing point (0°C), because of this large amount of heat is given up before water turn to ice and furthers due to restricted stirring oceans and lakes freezes only at the surface. The temperature of the medium therefore can drop only to 0°C in fresh water and lakes and few degree lower in ocean (up to –2.5°C).Between common compound and molecules, water is the only one whose solid form i.e. ice is lighter than its liquid form, it expends by about 8 percent when it freezes and become less dense. This is responsible for the floating of ice on water. Water density is 1.0 and water is heaviest at 4°C. In deep lakes and ponds the water do not freeze up to the bottom where the animal live comfortably.

Water is a unique liquid. Ahluwalia (2008) explains the properties of water as:

High dipole moment	=	1.84×10^{-3}
High dielectric constant	=	80
Specific heat	=	1Cal g^{-1} °C
High heat of evaporation	=	540 cal g^{-1}
Surface tension	=	73 dynes cm^{-1} at 20°C and
Viscosity	=	0.01 poise at 20°C

Above abnormal properties of water are due to hydrogen bonding. Natural water is slightly acidic (pH 5.6) due to dissolved carbon dioxide, which gives carbonic acid (H_2CO_3). Therefore it can dissolve a variety of compounds, from simple salts to minerals including sodium chloride and calcium carbonate in lime stone rock. Latent heat of water is responsible for moderating the temperature of the biosphere. It also plays a vital role in the evaporation of water and its condensation as rain and as dew. Water is poor thermal conductors as compared to other metals. Due to high viscosity of water, it allows organisms to swim using simple movement and, their light penetration power allows photosynthetic organism to live below the surface of water. Water exerts pressure; organisms living at sea level experience a pressure of about 15 psi (~ 1 atmosphere, which is equal to 360 mm mercury). With depth the pressure increases at the ratio of about one atmosphere for every 10 meter of decent. The physical characteristics like solubility, ionic dissociation and surface tension are all influenced by the pressure. In the language of chemistry, water molecules are a hybrid of oxygen and hydrogen. In water molecules two atoms of hydrogen share their electron with two unpaired electron of an oxygen atom, forming covalent bond and a stable molecule of H_2O is formed. Electronegativity of oxygen, creates a partial positive charge on each hydrogen atom and two partial negative charges on oxygen atom, due to this reason the water molecules become dipolar. The interaction of each pair of free electrons of oxygen atom of one water molecules can serve as an H acceptor to form hydrogen bond with a hydrogen atom of another molecule. Similarly each hydrogen atom of one water molecules can serve as an H donor to form hydrogen bond with an electron pair of oxygen atom of another water molecule. Thus each water molecules can link with four other water molecules by hydrogen bond. Since these hydrogen bond are non covalent i.e. these are not formed by sharing of electron, these are week bond yet, by forming a large number of hydrogen bond in every direction due to their strong intermolecular attraction. Water molecules form stable water with an extensive, three-dimensional lattice skeleton of hydrogen bond having a tendency to form tetrahedral structure. This lattice skeleton gives water its unique properties of fluidity, surface tension, greater viscosity than alcohol or chloroform, stronger cohesion and adhesion forces and high specific heat and latent heat of vaporization. If soluble salt like calcium, magnesium, chloride, sulphate and bicarbonate are present in water, that water is called hard water. It is two types:

(1) Temporary hard water

(2) Permanent hard water

Temporary hard water has calcium bicarbonate and magnesium bicarbonate. By boiling this hardness can be removed. The permanent hardness is due to the presence of chloride, sulphate of calcium and magnesium. Through chemical treatment permanent hardness can be removed. When the water has low salt content then it is called soft water. In soft fresh water, calcium and carbonate ions are relatively less in content than sodium and calcium.

Ragothaman and Trivedi (2002) have given a comparative study of soft, hard and seawater. (Table 1)

TABLE 1

Comparison of hard, soft and sea water

Water	Na	K	Ca	Mg	Cl	SO_3	CO_3	Total g/l
Soft water	0.016	–	0.010	–	0.019	0.107	0.012	0.065
Hard water	0.021	0.016	0.065	0.014	0.041	0.125	0.119	0.30
Sea water	10.7	0.38	0.42	1.31	19.3	2.69	0.073	34.9

In nature, water is found in four major forms.

(1) Atmospheric moisture

(2) Precipitation

(3) Surface water

(4) Soil water

Atmosphere moisture is present as invisible vapour, known as the humidity and also as visible vapour, known as fog or cloud. The invisible vapour or humidity of the air is usually expressed as relative humidity. Precipitation or rainfall is the chief source of moisture. Directly rain is of little important to plant. Soil moisture is chief source of water for planet. As surface water fresh water is present in rivers, lakes and streams. Water as an ecological factor occupied the foremost position in distribution of vegetation and its structure.

Warming (1909) classified plant communities on the basis of plant dependence upon and relation to water as done earlier by Grabner (1898, 1901, 1908). On the basis of their water requirements Warming primarily recognized three major group of plant.

(1) Aquatic plant (hydrophytes)

(2) Land or terrestrial plant

(3) Halophytes or marshy plant

The total volume of water in the hydrosphere is estimated to be about 1.36 x 10^9, of which 97.3% is locked up in the ocean as seawater and another 2.1% in ice caps and glaciers (Narasimhan, 2005). Freshwater, critical for sustenance of terrestrial life (plant, animal and human being) constitutes about 0.6% of the total water. A bulk of this freshwater occurs as a groundwater and as soil water, which can be extracted only by plant. Fresh water's in lakes, streams and the atmospheric constituent less than, 0.05 % of all water on the Earth. Atmosphere alone contain only about 13000 km^3 or 0.001 % of the Earth's water inventory.

The availability of such mobile water as contained in river, stream, groundwater, and the atmosphere, and small portion available for evaporation in the ocean, lakes and soil, is less than 0.05% of all the water available on the Earth. Yet, this small quantity of water, through vigorous cycling, has dictated the geological and biological history of the Earth for billion of years.

Current understandings of the hydrological cycle show that, this phenomenon involving a fraction of one percent of all water on the Earth sustains all life. The hydrological cycle is largely driven by solar energy. Of the total solar energy received on the earth surface about 40% is returned to the atmosphere as latent heat of evaporation and another 18% as sensible heat. The average annual precipitation over the entire surface of the Earth is estimated to be about 100 cm, amounting to a total volume of about 5 x 10^3 units this is about 39 times the quantity of all water in the atmosphere, implying that the average residence time of water in the atmosphere is for about 9.4 days. Rainwater contains about 8ppm of dissolve oxygen. The small amount of carbon dioxide in the atmosphere is dissolve in the rainwater to render it mildly acidic. The mild acidity and dissolve oxygen set the stage for chemical weathering of the lithosphere.

Huge quantity of water is retained in the soil. The amount varies from place to place. Rain is the main source of soil water. After the rain water is washed out through the slopes. This water is not available to the plant. This is called run away water.

The water that moves freely and reached the water table is called the gravitational water. The dry soil also contains a certain percentage of water. The thin film of water, which occurs over the soil particle, is called hygroscopic water. This water is not available to plants. The water that is readily absorbed by the root is called capillary water. The maximum water holding capacity is represented by capillary and hygroscopic water. The total amount of water retained by the soil, after the gravitational water has drained away is the field capacity of the soil. The water, which is absorbed by the plant, is termed as available water and one, which cannot be absorbed, is referred to as non-available water. The total water present in the soil is called holard. The available water which the plant can absorb is called chresard and the water which cannot be absorbed by the plants is called as echard.

IMPORTANCE OF WATER

The extensive tree dimensional lattice skeleton of water molecules makes water the most stable fluid on Earth.That is why water becomes the universal solvent in living system, forming both intracellular and extra cellular fluid matrices and constituting about 50% to 90% of the bodies of organism. No other earthly fluid compared with water is solvent of water. Evolution of all the diverse type of organisms from the most ancestral 'primordial organism' has been significantly shaped by properties of water. Being polar in nature, water molecules readily link electrostatically by weak hydrogen bonds with all types of ion and polar molecules. That is why, cellular poles comprises thousand of molelcules as solute in the cytosol. Due to this polarity of water molecules, water is a poor solvent for non-polar molecules like the lipids. This has been the basic of structure making in living system; the non-polar molecules from same membrane, making cell the basic membrane bound units of life and causing compartmentalization of cell architecture into organelles. Dynamic nature of water in an ideal dispersion medium with solutes in its diffuse uniformly, remain in constant motion that is Brownian motion, freely collides and readily interact. That is why, thousand of metabolic reaction occur in cells.Water flow freely in the cell, causing streaming (cyclosis) in cytosol. This facilitates exchange of water and many solute various cell organelles, as well as between the cell and their surrounding. Water play a crucial role in stabilizing the folded and coiled, three dimensional confirmation of various biopolymers that is polysaccharides, protein and nucleic acid. It also gives stability to the structural organization of the cell. Many enzymes driven metabolic reaction in cell require H^+ and OH^-. Since water is a weak electrolyte, some water molecules in cells are always undergoing self ionization (auto-ionization), producing H^+ and OH^- ions required by proteins, nucleic acid, phospholipids etc. for their specific roles in cell structure and function. Also, the mineral salts of cytosolic pool ionize in water and serve as cofactor in a number of enzymes. Water maintain a proper turger pressure in cells, i.e. turgidity, preventing cells from collapsing, helping cell to grow, and helping lower organism and soft parts of plant to maintain their shapes. Water is relevant in photosynthesis, contributing to the synthesis of food by green plant for all organisms. Remember, the oxygen present in the atmosphere and use by the organism in respiration is entirely a byproduct of photosynthetic and its source is solely the water, which participates in photosynthesis. Water is itself a reactant in many metabolic reactions. Conversely, water is a byproduct in all dehydration condensation reaction, which occurs in cells mainly for synthesis of biopolymers.

Due to strong cohesive and adhesive nature of water, water flows freely and continuously in vessels and duct, transporting nutrients, respiratory gases, carbon dioxide, oxygen, metabolic water, hormones etc. within the body. This property of water also helps in regulating exchange of material between body and external environment. Water is responsible for the turger pressure, which maintains body shape in small animals. For example, In plant it regulate closing and opening of stomata, buds and flower, coiling of tendrils, folding and unfolding of leaves, shriveling of leaves when touched as in *Mimosa pudica*. In animal, water regulates movement as those of starfishes by means of tube feet and of cnidarians by means of hallows tentacles. Because of surface tension and strong cohesive force, water creates capillary forces in this condition. This help in ascent of sap in plant from soil in root and stem also, it is because of surface tension and cohesion that we can fill bowls with water up to slightly above the brim of bowls and water-skating insects can even walk upon the surface.

Conservation water in the body and its bulk movements between body and environment help organism to adapt to various types of habitats such as deserts, marshes, ocean, freshwater bodies etc. Water is unique in that its solid state i.e. ice is less dense and more space occupying than liquid water. On freezing, therefore, ice starts forming from surface, and not from bottom of the water body and it spreads and keeps floating upon water surface. This surface sheet of ice insulated underlying water, limiting further freezing. Aquatic organisms depend on this property of water for their survival in winters. That is why water is the most ideal habitats for organism, especially animals hence; water has a rich aquatic fauna. Certain types of organisms survive in water even at Earth poles. Having a strong cohesion force due to extensive lattice framework of hydrogen bonds, water has a large heat capacity and hence, high melting point and boiling point. Cooling or warming of water, takes more time because it involves breaking and reforming of a large number of hydrogen bonds. That is why; large water bodies like ocean, river and lakes have moderating effect on environmental temperature of coastal area by warming cold air in winter and cooling warm air in summers. That is why, towns and cities along seacoasts and riverbanks have moderate temperature throughout the year. Further, the high specific heat and free flowing of water ensures uniform distribution of heat in the body. Water plays a crucial role in thermoregulation of body. It helps in regulation of body temperature by the way of heat loss from body surface by convocation (flow of large amount of body fluid near by surface) and radiation in both plant and animals. Further, water has an unusually high latent heat of vaporization. Vaporization of sweat from our skin surface, therefore, requires large quantity of body heat. That is how our body are cold for thermoregulation during summer. The unusually high latent heat of vaporization of water has a significant effect on Earth's weather. Water of our Earth absorbs about 30% of solar heat resulting into vaporization of surface water, forming cloud. Clouds burst at rains when water vapours condense, releasing back the solar energy of thunderclouds and hurricanes. Rains clear the atmosphere, cool the Earth and replenish the water bodies. This natural water cycle is called hydrological cycle of Earth

RESOURCES OF WATER

We know that about three fourth of the Earth is covered by water. Various resources of water, their present status and their future have been discussed by Kumar *at al.* (2005) in length. Our aquatic biome is divisible into marine and fresh water. A flow diagram of aquatic biome Dwivedi *et al.* (2007) is given in fig.1 which represents the.resources of water. The major aquatic ecosystem occurring on the Earth surface is referred as aquatic biome. The broad categories of aquatic ecosystem– fresh water, marine, estuarine- are distinguished on the basis of differences in their salt content. Composing more than 70% of the earth surface, aquatic ecosystem is not only the dominant feature of Earth but is also very diverse in species and complexity of interaction among the physical, chemical and biological component. Fresh water ecosystem covers about 2% of the Earth surface, an area of about 2.5 million km^2. Fresh water ecosystem is characterized as having running water that is lotic ecosystem or still water that is lentic ecosystem. Lentic water has been included all inland water in which the water has not been continually flowing in a definite direction. Lentic ecosystem that is pools, pond, swamps, bogs, lakes, marshes etc. vary considerably in physical, chemical and biological characteristics. The lotic environment is consisting of inland water in which the entire water body continually moves in definite direction. The river, stream, brooks, springs, rivulets, and lakes could be classified on the basis of depth and nutrient levels, as oligotrophic and eutrophic. Oligotrophic lakes have been regarded to be in its young stage, these lakes are generally very deep, with a sand or rock bottom, and have little nutrients. Eutrophic lakes have been regarded as middle aged and are relatively shallow in comparison to the former, they are with silty or mud bottom and have sufficient nutrients.

The major zones in marine ecosystem are the littoral, neritic, pelagic and benthic. The littoral or intertidal zone is the shoreline between land and open sea. The neritic zone is by a continental shelf, extending outward to its edge to a water depth of about 200 m. This zone constitutes about 7.5 % of the total ocean area and is relatively rich in species and high in productivity owing to the depth of light penetration and the presence of nutrients washed from the land. This richness and productivity decrease gradually towards the deeper water of the zones. The pelagic zones of the open sea is constituting about, 90% of the total ocean surface. Photosynthesis occurs in the surface of pelagic zones primarily by planktonic diatoms and dianoflagellates. The benthic zone extends from the edge of the continental shelf to the deepest ocean trenches. Estuary is an ecosystem in which freshwater from river meets ocean water, the two being mixed by action of the tides. Estuaries water includes shallow bays, tidal, marshes and bodies of water behind barrier beach.

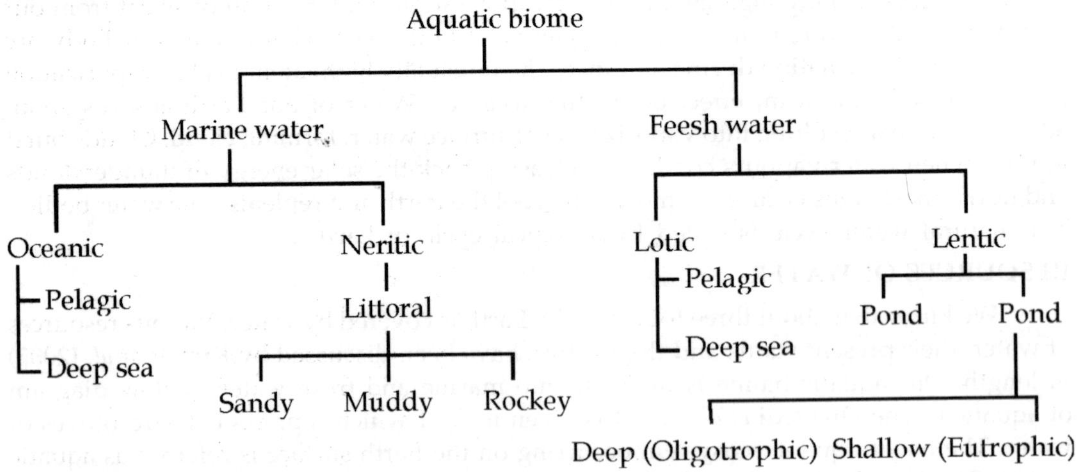

Fig: 1 : Aquatic biome

RESOURCES OF WATER IN INDIA

Country like India is still largely dependent upon rain for agriculture. Availability of freshwater is one of foremost concern for the future. Most of Indian plains receive about 80 % of there annual rainfall quota of rain from the south west monsoon during the four months, June to September (Pant *et al.*, 1997). India receives annual precipitation of about

4000 km^3. The total average annual flow per year for the Indian river is estimated as 1953 km^3. The monsoon rainfall is of the order of 3000 km^3. Rainfall in India is dependent on the south-west and north-east monsoon, on shallow cyclonic depression and disturbance on local storms. Lal (2001) mentioned that more than 50% of water resources of India are located in various tributaries of river like Ganga, Brahmputra and Indus river system. The highest rainfall in India of about 11690 mm is recorded at Mousinram near Cherrapunji in Meghalaya in the north-east (Bose *et al.*, 2001). Normal duration of monsoon in India is about 100 to 120 days beginning from first June. In India the two monsoon seasons, the south-west monsoon in June to September and the north-east monsoon in November-December bring rain (Table 2, Indian Agriculture in Brief, 1970) The annual potential natural groundwater recharge from rainfall in India is about 342.43 km^3, which is 8.56% of the total annual rainfall of the country.

TABLE 2

Season-wise distribution of rain fall in India

Rainfall seasons	Duration	Approximate % of annual rainfall
South-west monsoon	June-Sept	73.3
Post- monsoon	Oct-Dec	13.3
Winter or North-East monsoon	Jan-Feb	2.6
Pre- monsoon	March- May	10.4

POLLUTION OF RIVER IN INDIA AND ABROAD

Rivers in India have been regarded from time immemorial as sacred water bodies. It is believed that a holy dip in the same purges away all sins. Hence, million of people in India take holy dips in river, especially on some auspicious occasion. India is a land of many rivers. Importance of river is universally known since ancient period. Most of the cities are situated at the bank of one or other river. Recently many river have lost there identity due to pollution. Pollution which is produced from industries i.e. dyeing, textiles, paper, and pulp mills, battery, dairy, tanneries, breweries, steel industries, mining operation, agricultural practices i.e. artificial fertilizer, pesticides and biocides, sewage garbage, heavy metals etc. are directly discharges into river which contribute to pollution of river. Owing to negligence or ignorance, we fail to notice that during process of progress and development we have ruined our natural heritage, the prominent water bodies, our rivers, we dump our entire waste in those river which carry them away from us. At the same time we depend upon the river directly or indirectly for activities other than mentioned above. It may before fishes, drinking water, recreation etc. if the quality of water is spoiled than the other products are also malformed. The fishes may show biological magnification or the water we drink may be polluted. In most of the cities, the municipal water supply is directly pumped up from the river of the city. Thus, the importance of water quality of a river can not be ignored; further all effort be made to control the water quality.

WATER STANDARD

Standards are necessary for controlling the polluted water resources so that it would continue to be suitable for man and other species of the aquatic system. No single parameter is possible to determine the quality of water. Depending on the use for which water is intended to be used, it should have the right parameter.

Standard of different category of water have been prescribed by different health agencies. Some such types of agencies are:

USPHS = United State Public Health service

ICMR = Indian Council of Medical Research

WHO = World Health Organization

ISI = International System of Units

UPPCB = Uttar Pradesh Pollution Control Board

CPCB = Central Pollution Control Board

BIS = Bureau of Indian Standard 1991

Water quality standard of some agencies is given in table 3.

TABLE 3

Standard of water quality

Parameters	BIS	ISI	UPPCB	USPHS	WHO	ICMR
Clour	Colourless	—	—	—	—	—
Odour	Odurless	Unobjectable	—	—	—	Unobjectable
Temperature (°C)	—	40	—	—	20 – 22	—
pH	6.5 – 8.5	6.0 – 9.0	5.5 -9.0	6.0 -8.5	7.0 – 8.5	6.5 – 9.2
Turbidity (NTU)	—	—	—	—	5	—
Conductivity (mho)	—	—	—	—	300	—
T.S.(mg/l)	—	—	—	500	500	—
T.D.S. (mg/l)	500	500	—	—	—	—
T.S.S. (mg/l)	—	100	—	—	—	—
D.O (mg/l)	4.6	>5.0	—	>4.0	—	—
B.O.D. (mg/l)	—	30	30	—	—	—
C.O.D. (mg/l)	—	<20	250	—	—	—
Hardness (mg/l)	—	300	—	500	100	300
Free CO_2 (mg/l)	—	—	—	—	—	—
Chloride (mg/l)	250	250	600	250	200	250

IMPACT OF WATER POLLUTION

When quality and composition of water is altered by anthropogenic interference, such that it no more be safe for direct or indirect consumption then the water is said to be polluted. Consumption of polluted water may lead to various abnormalities, which may or may not be visible to us. Through the process of biological magnification the heavy metals and unwanted organics present in polluted water ultimately reach to human either through flesh, milk or grains (Dwivedi *et al.*, 2006). Their accumulation in vital organs not only leads to death of the individuals but also to various abnormalities, such as infertility, mental disorder or many new symptoms.

(A) Effect on Ecosystem

Oxygen demanding waste, like sewage contains large amount of organic material that can be oxidized by microorganisms. Degradation of organic compounds leads to O_2 depletion in the water bodies which effect or even kills aquatic life forms. Runoff from agricultural fields carry nitrogenous and phosphate nutrients into the water, which increases the growth of aquatic plants. Later, often their death followed by decomposition causes blooms and eutrophication. The waste produced by industries like dairy, tannery, distillery, oil refinery, textile, coal and synthetic rubber etc are highly toxic. Inorganic and synthetic organic pollutant cause several adverse effects on the life of the aquatic system. Plants as well as the animal life both are affected by water pollution. Sensitive plants, fishes as well the microorganisms in the affected area are displaced breaking the food chain in the area. As a result when the plants die and decompose, create marshy habitat, which not only reduces the oxygen content, but it also creates eutrophic condition. Use of polluted water for irrigation is in practice in sub-urbs which affect the properties of the soil and also the microbial population in the soil. Heavy metals also make entry into the food chain through this way.

(B) Problem for Health

Polluted water is a potential source of infective bacteria and other microbes which act as infective agent causing disease in man and other animal, for example, cholera, typhoid, dysentery, hepatitis, gastroenteritis, snail fever, enteric fever, malaria and many skin diseases. Polluted water may have organic chemicals like detergents, pesticides and chlorinates compound such as D.D.T. (di chlro di phenyl tri chlro ethane), BHC (Benzene hexa chloride), PCBS (Polychlorinated biphenyls) and benzene compound. These are hazardous compound, which may cause extinction of the keystone species, disturbing the complete ecosystem (Dwivedi and Shashi, 2006). Many inorganic mineral and compounds like lead, mercury, cadamium, copper, cyanides, acid etc are toxic for aquatic life and human beings. For example Lead affects the central nervous system reducing fertility and mental retardation in children. Mercury causes minamata disease, Arsenic and Cadmium produces dermal problem, stomach pain, vomiting or diarrhea. Fluorides cause dental fluorosis, pain in bone and joints. Radioactive materials cause several genetic diseases, which can be very serious problem of human life. Oil film, which spreads over the surface of water, prevents oxygenation of water. It inhibits planktonic growth and photosynthetic activity of aquatic life.

MANAGEMENT OF WATER

Under the pressure of increasing population growth we feel ourselves dwarf in controlling the water pollution and therefore, we are left behind only with the solution of proper management of the water bodies. We are trying to manage our water resources and apply some treatment system for sanitation of water as discussed below:

(A) Physical Treatment

The objective of physical management is to remove the gross solids through the process of screening and floatation

(a) Screening

By this method suspended large particulate matters are separated mechanically by using static vibration and rotating screens.

(b) Floatation

Floatation is applicable when the wastewater is collected in a large basin for some time. In the basin air bubbles cause the suspended solid particles to adsorb them and help to rise up to the surface of basin, where these are removed by mechanical means. Floatation helps in the removal of lighter matters like grease, oil, fibers, coagulants, clothes, wood etc.

(B) Chemical Management

The chemical management for the polluted water can be divided into three steps.

(a) Primary step

This step is useful in removing the remaining suspended solids, through the process of sedimentation, precipitation or neutralization. In sedimentation, the polluted waste water is sedimented in large tank physically or adding some coagulants like lime, hydrated aluminiums, sulphate, ferric salts etc. These chemicals, in solution furnish ions contrary to those of the colloidal particles in the waste water, producing jelly like spongy mass of suspended matters. This suspended matter is passed through a tank fitted with paddles, rotating gently. As a result the fine particles clump into larger particle and settle down. For maintaining pH, the acidic water is neutralized by adding lime stone or caustic soda while alkaline water is neutralized by adding H_2SO_4 and CO_2.

(b) Secondary step

In secondary step, the dissolved and colloidal organic matters present in waste water is removed by biological processes using microorganisms. These aerobic and anaerobic microorganisms cause biodegradation of organic components in the polluted water. The polluted water from sedimentation tanks is first subjected to aerobic oxidation system, such as aerated lagoons, trickling fitters, activated sludge units, oxidation ponds or ditches. In these, almost all organic substances (except hydrocarbons and ethers) can be oxidized by aerobic biological treatment. The sludge obtained is subjected to anaerobic digestion in the sludge digesters and water is either used for irrigation purpose or for further processing.

(c) Tertiary step

This step is used for the removal of suspended solid, bacteria etc, through the process of reverse osmosis, electro dialysis, ionic exchange, adsorption etc. Reverse osmosis is done by passing the waste water through the semipermeable membrane of cellulose to separate out phenol, glycol and alcohols from waste waters. Electrodialysis is done by creating electric potential between two electrodes, thus the substances are filtered by using an ion selective membrane ahead to the electrodes. Removal of (Cu) cupper, (Pb) lead, (Hg) mercury, (Cr) chromium), (Ni) nickel etc. from the wastewater is done by synthetic ion exchangers (Cation exchanger and anion exchanger) and resin. Activated carbon is also generally used as absorbent for removal of toxic heavy metal ions, D.D.T. and carbonate insecticide from the waste water. Then finally the waste water is chlorinated to kill the microorganisms. Further, chemical disinfections may also be done.

(C) Management of Water by Aquatic Plant

Many aquatic plants are potent to remove pollutant from wastewater. Certain plants are the good indicator of water pollution for example *Utricularia, Chara, Wolfia* and Algae like *Cladophora, Chlorella,* etc. An extensive list of such plant has been prepared by Das and Datta (2006). Phytoremediation is a process in which recovery of hazardous substances from soil or ground water contaminated with municipal or industrial wastes etc. is done using plants. The water hyacinth can absorb heavy metals like Pb (lead), Cd (cadmium), Hg (mercury), Ni (nickel), and phenolic compound together with Nitrogen, Phosphorus, Potassium, Sulphur and Calcium (Upadhyay, 2004). All these can be removed from waste water by periodic harvesting of hyacinth biomass from wastewater channels. Alage like *Microcystis, Chlorella, Chlamydomonas,* and *spirogyra* are reported to be potent to accumulate radioactive substances, heavy metal, Nitrate, Sulphur, organophosphorus, aromatic hydrocarbons, etc. These chemicals can be removed from polluted water through the harvesting of algal biomass, Sadgir *et al.,* (2005) found that *Phyllanthus emblica* is an effective as coagulant and can be used coagulant aid in the water purification system. According to Nafisha *et al.* (2004), if water hyacinth is treated with lignolytic fungi (*Phanerochaete chrysosporium*) and chemical (Pyrogollol red) they produce sufficient biogas. So those aquatic plants, which have rapid growth rate, absorb large quantities of nutrients and they might provide a practical and economical method for management of water pollution.

CONCLUSION

Environmental resources like the atmosphere and water may be regarded as global public goods. Their conservation not only benefits the local population but also the fauna and flora of the region. Owing to negligence we fail to notice that during the process and development we have ruined our natural heritage, the prominent water bodies; our rivers. Water is polluted by many kinds of substance i.e. traditional organic matter, waste generated from industrial processes, chemical agent of fertilizers and pesticides used for crop protection and silt from degraded catchments. While it is estimated that three forth volume of the waste water is generated from municipal sources; industrial wastes contributes over one half of the total pollution load and major portion of this is coming from large and medium industries (Dwivedi, *et al.,* 2009). The 17 industries declared as the most polluting are sugar, fertilizer, cement, fermentation and distillery, aluminium, petrochemical, thermal power, caustic soda, oil refineries, tanneries, copper smelters, Zn smelters, iron and steel, pulp and paper, dye and dye intermediates, pesticides and pharmaceuticals. The figure released by the ministry of environment and forest on March 30, 1993 shows a total of 1629 unit (Martin, 1998) in India, which fall under this category's. Generally, the pollutant come from sewage discharge, industrial effluents discharged without any pretreatment and surface runoff from agricultural land where chemical fertilizers, pesticides, insecticides and manures are used. This makes the river water unsafe for drinking and bathing, which needs to be prevented, for the present as well as the future generations.

REFERENCES

Ahluwalia, V.K. 2008. Environmental chemistry. Ane Book India, New Delhi : 30-31.

Balaram, P. 2005. Editorial issue. *Current science* 89 (5):732.

Law, B.C. 1972. Historical geography of Anicient India. Publised by UP Hindi Granth Academy, Lucknow: 111.

Bose, D. and Saxena, N. 2001. Monsoon. Employment News, 2001, xxvi No.8

Das, G.K. and Datta, S. 2006. Managing waters of wetlands in and around Kolkata. *Indian Science Cruiser,* 20 (3): 22.

Dwivedi, A.K., Hasan, H. and Shashi 2009. Anthropecology of Water Bodies, In Plant Physiology in Agriculture and Forestry (Ed. P.C. Trivedi), Aavishkar Publishers and Distributors, Jaipur, India: 248-254.

Dwivedi, A.K., Prajapati, U.B. and Shashi 2007. Pollution of River Aami in India.http://www.shvoong.com/exact-sciences/earth-sciences/water-science/1687304-pollution-river-aami-india/

Dwivedi, A.K. and Shashi 2007. Species erosion, the shrinking biodiversity- An approach, *Indian Science Cruiser,* 20(5):8-14.

Dwivedi, A.K., Singh, R. and Shashi (2008) Grondwater Pollution: Causes and Impact, *Indian Science Cruiser,* 22(4): 29-33.

Dwivedi, A.K., Shashi and Singh, J. 2006. Water pollution and groundwater recharging. *Current Science,* 91(4): 407-408.

Dwivedi, A.K., Prajapati, U.B. and Shashi 2007. Waste Water and its Management, *Indian Science Cruiser,* 21(4): 36-40.

Karmondy, E.J. 2002. Concept of Ecology, Prentice-Hall of India, Private Limited, New Delhi 110001.

Gupta, R. 2007. A text book of modern zoology Prakash publication, 46 Zila parishad market,Muzaffarnagar 251001 (U P) India: 103.

Indian Agiculture in brief 1970. directorate of Economics and statistics, Government of India, New Delhi) 10th edn.:18.

Karthikeyan, K. and Singh, K. 2004. Chemical composition of some sewage water used for irrigation purpose. *J. of industrial pollution control* 20(2): 211-219.

Kumar, R., Singh, R.D. and Sharma, K.D. 2005. Water resources of India. *Current Science,* 89 (5): 794.

Lal, M. 2001. Climate change- implication for india's water resources. *J. India water Res Soc.* 21:101-119.

Mahajan, S.P. 1985. Pollution Control in Process Industries. Tata Mc Graw Hill. New Delhi: 289.

Nafisha, Ali, Chaudhary, B.L. and Khandelwal, S.K. 2004. Better use of water hyacinth for fuel manure and pollution free environment. *Indian j envir prot,* 24 (4): 297.

Narsimham, T.N. 2005. Water: Science and Society. *Current Science* 89 (5): 786-793.

Pandey, Y.N. and Rao, G.P. 2000. Introduction to Applied Ecology. Student Friends, Allahabad.

Pant, G.B. and Kumar, R. 1997. Climates of South Asia, John Wiley and Sons, Chiche sterl. : 320.

Ragothaman, G. and Trivedi, R. K. 2002. Aquatic Ecology. Agrobios (India), Jodhpur : 25.

Rajurakar, N.S., Nongsri, B., and Wardhan, A.M. 2003. Physico-chemical and biological investigation of river Umshyrpi at Shillong Meghalaya. *Indian J envir Prot.* 23(9): 990-998.

Right to water is fundamental 2007. Yojana. September : 3.

Sadgir, P., Garg, Patel, A.N., Bhan-oarir and Sharmak 2005. Activity of amla or *Phyllanthus emblica* as coagulant and coagulant aid for water purification. *Res j chem. Envir,* 9 (1): 49.

Singh, A.P. and Singh, J. 2003. Physico-chemical characteristics of river in relation to discharged of paper mill effluent. Global Science Academy, Basti, *Indian J Environ Health* April 45 (2): 93-96.

Upadhyay, A.R. 2004. Aquatic plant for wastewater treatment, Daya Publishing House. Delhi.

□□□

Chapter 5

Assessment of Quaternary Vegetation and Climate of Kumaun Himalaya (Sukha Tal)—Geochemical Analysis

ASHA GUPTA

INTRODUCTION

International efforts are afoot to develop more proxy climate data to understand intricate mechanism regulating earth's climate. Himalaya – a world's highest mountainous chain, forming the northern part of India, is tectonically highly disturbed but has required potential to yield the needed data of great use. The knowledge of past vegetation has been considered a most reliable parameter for the interpretation of past climate. Quaternary sediments of Indian Himalayan region are pollen analyzed since beginning of nineteenth century to reconstruct past vegetation and understand the climatic changes in the recent past and have covered different regions. Kumaun Himalaya (lying between latitude 28°44' - 31°25' N and longitude 78°45'- 81°1' E) is studied since over 4 decade and earlier workers have sufficiently covered its subtropical zone (Mittre *et al* 1967, Gupta 1977, Gupta and Khandelwal 1982, Chauhan and Sharma 1996, Kotlia *et al* 1997, etc.). Recently, the study has been extended to its temperate zone and beside conventionally used pollen analysis (Gupta 2002a, 2007a, 2008a), the geochemical and palaeontological analyses of same samples (Gupta 2003a, b, 2004a, 2005a, b, 2006a, b, 2007b, 2008b,c) have also been attempted, in order to develop more authenticity about the generated data and such studies have opened concealed neotectonic disturbances (Gupta 2002b, 2006c, 2009).

Present contribution provides the account of geochemical analysis of sedimentary profile from Sukha Tal which has revealed the knowledge of total organic matter, carbonate contents and corresponding climate of the region since Early Holocene. The detailed account of pollen analysis of same profile has already been published separately (Gupta 2008a) while of Paleontological analysis be available elsewhere. These studies have revealed presence of concealed fault at this site (Gupta 2006c) and due to this fault a

Birbal Sahni Institute of Palaeobotany, 53, University Road, Lucknow - 226 007, India.

part of sequence (near onset of Late Holocene) is represented two times. Geochemical data has been compared to the data of pollen analysis which has provided not only supportive but additional information about the reconstructed past vegetation.

Sukha Tal is situated in Naini Tal District, at 1981 m a.s.l., in the west-north-west of Naini Tal lake, ca. 3 km from its upper part Malli Tal. It is a dry lake and the lake area is demarcated by depression which is ca. 200 meters long and ca.100 meters broad. Originally, it is a part of Naini Tal lake which became separate latter on. The modern vegetation of Kumaun is reported frequently (Duthie 1918, Champion 1923, 1936, Osmaston 1927, Jain 1956, Bhargava & Gupta 1958, Gupta 1968, Champion & Seth 1968, Pandey *et al* 1984, etc.). However, area around investigated site shows mixed oak forest with mainly *Quercus incana* followed by *Rhododendron arboreum*. The climate of Naini Tal region is monsoonic temperate, having annual average rain fall of about 279.4 cm. The temperature varies 02.8°C – 15.6°C during winter and 10.6°C – 26.7°C during summer.

Geological details of Kumaun are well documented (Middlemiss 1890, Tiwari & Mehdi 1964, Pande 1974, Fuchs & Sinha 1974, Hukku *et al* 1974, Pal & Merh 1974, 1975, Raina & Dungrakoti 1975, Valdiya 1980, 1988, Valdiya *et al* 1984, etc.). The temperate zone of Kumaun lies in southern part of Lesser Himalaya which is bordered by Main Boundary Thrust. The latter is tectonically active and caused tremendous natural disturbances. Such disturbances are visible all around Sukha Tal (Figure-1). Presence of concealed fault at Sukha Tal matches to geological conditions of the area. Presently investigated profile represents both arms (blocks) of this fault, though the upper one is covered completely while lower by upper part only.

MATERIAL AND METHODS

Presently investigated profile was collected by the author from the surface of Sukha Tal lake, by trenching, during April, 2000 and maximum depth could cover was 3.5 metres because below which water seeped in. The sampling was done at an interval of 5 cm each. Sediments are largely comprised mainly of silty-clay with or without sand, granules, pebbles (breccia as well as conglomerate type). The lithological succession is shown in Table 1.

TABLE 1

Lithological succession of investigated profile

Unit	Depth (cm)	Description
1	0- 42	Silty-clay with sand
2	42-70	Silty-clay with sand and granules (conglomerate type)
3	70-200	Silty-clay with or without sand
4	200-240	Silty-clay with sand, granules and fine pebbles (conglomerate type)
5	240-350	Silty-clay with sand, granules and fine-coarse pebbles (breccia type)

About 20 grams material of each sample was used for the study. Initially all samples were completely dried in the oven, then kept in the muffle furnace for ignition at specific temperatures (i.e. at 550ºC and subsequently at 950ºC, for two hours each). The organic matter and carbonate contents were estimated on the basis of weight loss on ignition – a conventional method (Bengston & Enell 1986, Mooney 1997, Heiri *et al.*, 2001, etc.).

RADIOCARBON DATING

Five samples of the investigated profile have been radiocarbon dated, in the Radiocarbon Laboratory of Birbal Sahni Institute of Palaeobotany, Lucknow. Their details are as follows –

Sample 1: Depth 35-40 cm, Registered No. BS-2555, Age 2670 ± 110 years B.P.

Sample 2 : Depth 125-130 cm, Registered No. BS-1989, Age 6120 ± 170 years B.P.

Sample 3 : Depth 230-235 cm, Registered No. BS-1771, Age 8260 ± 170 years B.P.

Sample 4 : Depth 280-285 cm, Registered No. BS-1835, Age 2690 ± 90 years B.P.

Sample 5 : Depth 335-340 cm, Registered No. BS-2499, Age 3790 ± 110 years B.P.

The radiocarbon dates are not in chronological order, so profile in question is divided in to two units – upper (above 260cm) and lower (below 260 cm).

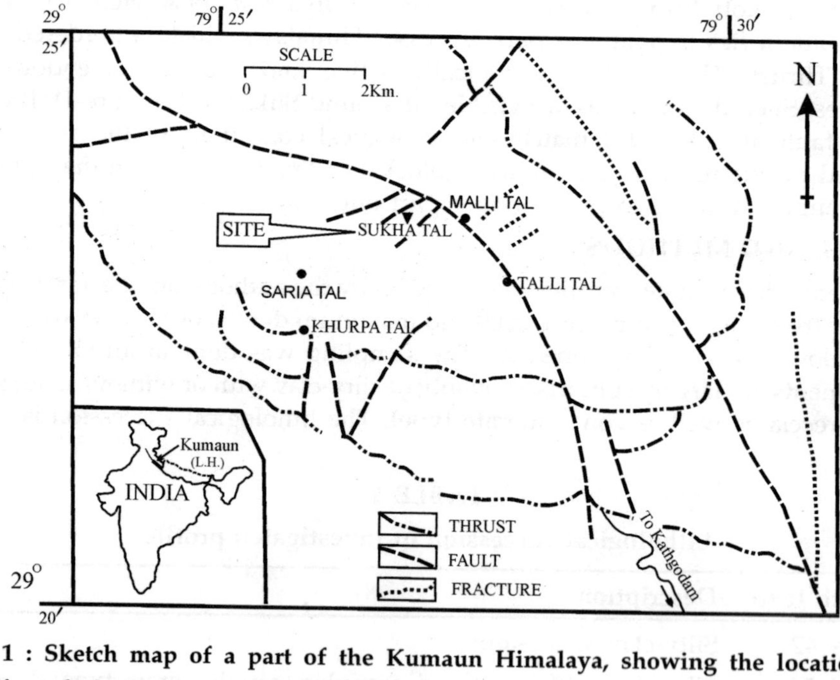

Fig. 1 : Sketch map of a part of the Kumaun Himalaya, showing the location of investigated site Sukha Tal and tectonic disturbances around it (after Valdiya 1988).

OBSERVATION

All samples have loose their weight on ignition at the applied temperatures. The difference in their weight prier and after such treatments has revealed the knowledge of total organic matter and total carbonate contents, which are plotted in the Figure-2. A part of profile i.e. 275-245 cm, has been discarded from the analysis, to avoid contamination of sediments of the upper and lower units. In order to get precision in evaluating the knowledge of past, the upper unit of profile has been divided in to 3 zones which are numbered I, II and III in ascending chronological order, prefixed after site abbreviation ST. The lower unit coincide with lower part of zone ST-III of upper unit, thus named zone ST-IIIa. The geochemical zones of both the units are described separately.

Upper Unit

Zone ST-I (Depth 245 – 200 cm)

A radiocarbon date i.e. 8260 ± 170 years B.P. is obtained from lower part (i.e. 230-235 cm depth) of the zone. This zone is characterized by having well representation of total organic matter (3.1 - 5.9 %) and low total carbonate contents (3.1 - 4.2 %). Such scenario indicates humid climate at the region.

Zone ST-II (Depth 200 – 50 cm)

Again only one ^{14}C date is available, which is 6120 ± 170 years B.P., from near middle part (125-130 cm) of the zone. This zone shows mark decrease of total organic matter (1.3–2.2 %) which is highly reduced at around 4200 years B.P. The frequency of total carbon contents shows abrupt increase (8.1 – 11.1 %). The change in the representation of organic matter and carbonate contents reflects change in climate towards moist (arid) conditions.

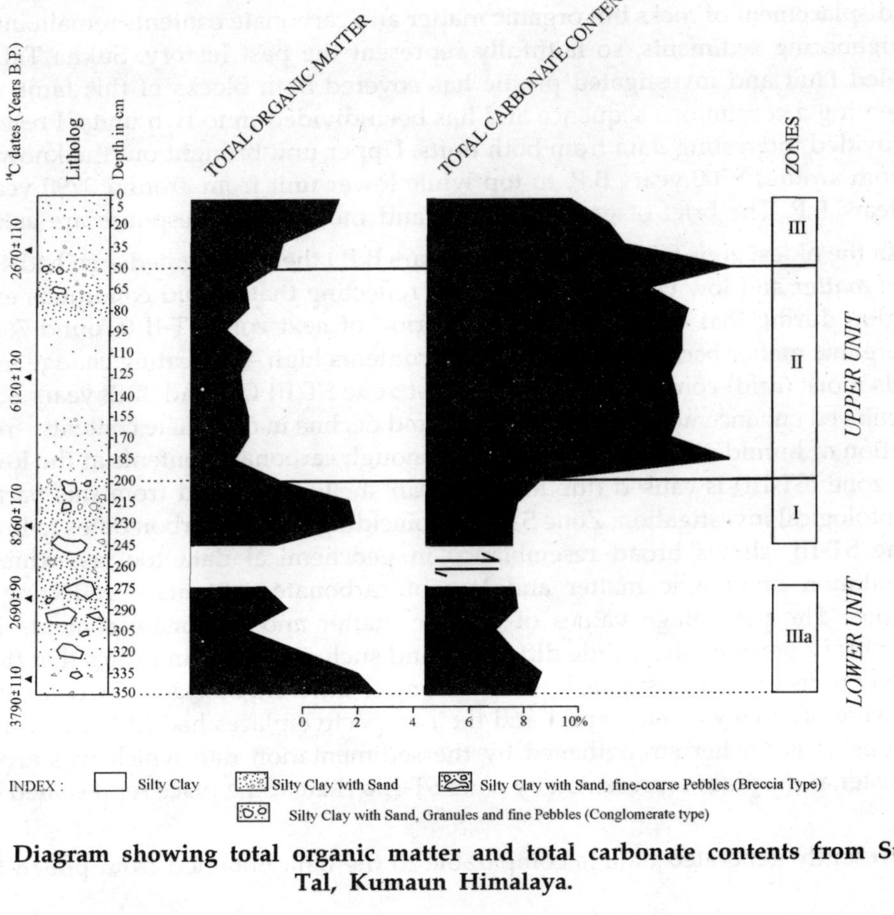

Fig. 2 : Diagram showing total organic matter and total carbonate contents from Sukha Tal, Kumaun Himalaya.

Zone ST-III **(Depth 50 – 0 cm)**

The lower part of this zone (35-40 cm) is radiometrically dated to 2670 ± 110 years B.P. This zone shows recoup of total organic matter which again became high (3.1–5.3 %). Total carbonate contents showing high value in the beginning became considerably declined (from 11.1–5.2 %). Such change in the representation of organic matter and carbonate contents reflects restoration of humid climate at the region.

Lower Unit

Zone ST-IIIa **(Depth 350 – 275 cm)**

Two radiocarbon dates are available – one each from lower and upper part of the zone. The former part (i.e. from 335-340 cm depth) is dated 3790 ± 110 years B.P. and latter (i.e. from 280-285 cm) 2690 ± 90 years B.P. This zone shows high representation of total organic matter (2.1 – 6.7 %) and low of total carbonate contents (2.5 – 4.2 %) which reflects humid conditions at the region.

DISCUSSION

The collision between Indian and Asian plates is causing not only upheaval of Himalaya but also tremendous tectonic disturbances at the region. Such disturbances cause displacement of rocks but organic matter and carbonate contents remain intact with the neighboring sediments, so faithfully represent the past history. Sukha Tal consists concealed fault and investigated profile has covered both blocks of this fault, thus not representing a continuous sequence and has been divided in to two units. Present study has provided interesting data from both units. Upper unit brought out the knowledge of past from around 8700 years B.P. to top while lower unit from around 3790 years B.P.– 2690 years B.P. The brief of generated data and their climate response are as follows–

In the oldest zone ST-I, (around 8700 years B.P.) the investigated area had sufficient organic matter and low carbonate contents – reflecting that humid conditions existed at the region during that tenure. During the period of next zone ST-II (around 7600 years B.P.), organic matter became low and carbon contents high – indicating change in climate towards moist (arid) conditions. In upper most zone ST-III (around 3713 years B.P.), area again enjoyed enhancement of organic matter and decline in carbonate contents – reflecting restoration of humid conditions. Presence of enough carbonate contents at the lower part of this zone (ST-III) is caused due to molluscan shells, recovered from this part during palaeontological investigation. Zone ST-IIIa, coinciding in radiocarbon dates to lower part of Zone ST-III, shows broad resemblance in geochemical data too, as exhibits well representation of organic matter and low of carbonate contents – reflecting humid conditions. The percentage values of organic matter and carbonate contents in these contemporary portions show little difference and such difference indicates that the upper and lower units of the investigated profile belong to different blocks of the concealed fault which were originally located apart and their respective places had little dissimilar local conditions. It is further strengthened by the sedimentation rate which was around six times faster at the place represented by zone ST-IIIa than at the place represented by zone ST-III.

Presently generated data is comparable to the data obtained from pollen analysis of same profile and has opened additional information about the reconstructed past vegetation. Pollen analysis revealed that in Early Holocene the study area had a predominance of non-arboreal with poor presence of arboreals. Soonafter, the latter expanded and resulted mixed oak forests at the region, which became considerably declined around 4200 years B.P. Near the onset of Late Holocene, these forests restored with amelioration of climate. Geochemical analysis has shown high representation of organic matter in the beginning (in its zone ST-I) which indicates that non-arboreals not only predominated over arboreals but also densely studded the investigated area. Besides, low representation of organic matter in the next zone (i.e. zone ST-II) reflects that the established mixed oak forests were not very dense. The decline in mixed-oak forest around 4200 years B.P. is strengthened, as geochemical study also shows decline in organic matter during this time. The upper most zone (i.e. zone ST-III) shows increase in organic matter and decrease in carbonate contents – supporting restoration of forest with amelioration of climatic conditions. Besides, the contemporary portions (i.e. lower part of zone ST-III and zone ST-IIIa) are also comparable like those of pollen record and further strengthens the generated data and their climate response at the region.

The climatic inferences drawn from present study broadly corroborates to the global climatic events. During climatic optimum (recorded between 8000–6000 years B.P.), the investigated area had well representation of organic matter with humid climate. Around 4200 years B.P. there was decline of organic matter – reflecting forest clearance (due to abrupt shift to cooling). During Medieval warming (recorded between 740–150 years A.D. i.e. ca 1210–800 years B.P.) this region had sufficient organic matter with humid climate. Besides, expansion of anthropogenic activities (ca 200 years B.P. onwards) may also be considered reflected, as show further enhancement of organic matter during this tenure.

REFERENCES

Bengtsson, L. & Enell, M. 1986. Chemical analysis. *In* : Handbook of Holocene Paleoecology and Paleohydrology (Ed. B.E. Burgland). John Wiley and Sons Ltd. : 423-451.

Bhargava, K.S. & Gupta, R.K. 1958. Seasonal Material for a Flora of Nainital. *Agra Univ. Jour. Research : Science* 7 : 1-48.

Champion, H.G. 1923. The influence of the hand of man on the distribution of forest types in the Kumaun Himalaya. *Indian Forester* 49 : 116-136.

Champion, H.G. 1936. A preliminary survey of the forest types of India and Burma. *Indian Forest Records (n.s.) Silviculture* 1 : 1-236.

Champion, H.G. & Seth, S.K. 1968. *A Revised Survey of the Forest Types of India.* Managers of Publications, Delhi.

Chauhan, M.S., & Sharma, C. 1996. Pollen Analysis of Mid-Holocene sediments from Kumaun Himalaya. *Proc. Symp. NW Himalaya Foredeep,* (GSI Spl. Publ.) : 257-269.

Duthie, J.F. 1918. *Catalogue of the plants of Kumaun and of the adjacent portions of Garhwal and Tibet, based on the collections made by Strachey and Winterbottom during the year 1846 to 1849 and on the catalogue originally prepared in 1852 by Lt.- Genl. Sir Richard Strachey revised and supplemented by J.F. Duthie.* Dehra Dun / Delhi.

Fuchs, G. & Sinha, A.K. 1974. On the Geology of Naini Tal (Kumaun Himalaya). *Him. Geol.* 4 : 563-580.

Gupta, A. 2002a. Palaeovegetation and Past Climate of Late Holocene from Temperate zone in Naini Tal District, Kumaun Himalaya. *Acta Palaeontologica Sinica* 41 : 517-523.

Gupta, A. 2002b. Research Finding : Biocontents – Useful to trace Tectonic disturbance. *BSIP Newsletter* : 5.

Gupta, A. 2003a. Palaeontological Investigation of Quaternary sediments from Himalaya and its Significance for Stratigraphy and Pollen Analysis. *XIX Indian Colloq. Micropalaeontol. Stratigr. Abst.* : 60.

Gupta, A. 2003b. Biodiversity in Temperate Zone of Kumaun Himalaya during Holocene – A Palynological and Palaeontological Assay. *Nat. Conf. Biodiversity Applied Biol. Plants Abstr.*: 58-59.

Gupta, A. 2004a. Holocene Molluscs from Kumaun Himalaya. *Proc. 91st Indian Science Congress (Earth System Sciences)* : 44.

Gupta, A. 2005a. Holocene Organic Matter and Climate at Sukha Tal, Kumaun Himalaya – A Geochemical Analysis. *Nat. Conf. Advancing Frontiers Biotechnol. Abstr.* : 20.

Gupta, A. 2005b. Climate Assessment of Himalaya – Multidisciplinary Approach. *Diamond Jubilee National Conference Challenges in Indian Palaeobiology – Current Status, Recent Developments and Future Directions Abstr.* : 44.

Gupta, A. 2006a. Holocene Molluscs from Saria Tal, Kumaun Himalaya. *BSIP Newsletter*, no. 9 : 11.

Gupta, A. 2006b. Multiproxy Analysis of Quaternary deposits from Kumaun Himalaya – An assessment of Past Organic Matter and Climate. *Diamond Jubilee International Conf. Changing Scenario Palaeobot. Allied Subjects Abstr.* : 50-51.

Gupta, A. 2006c. Past Organic-Matter and Neotectonic disturbance in Himalaya. *BSIP Newsletter*, no. 9 : 11.

Gupta, A. 2007a. Palaeovegetation and Past Climate since Middle Holocene in Temperate zone of Kumaun Himalaya, India (with remark on natural disturbances at the region). *Chenia* 9 : 173-187.

Gupta, A. 2007b. Assessment of Quaternary Vegetation and Climate of Himalaya (Saria Tal)–Geochemical Analysis. *XXX Annual Conference of Indian Botanical Society Abstr.* : 125.

Gupta, A. 2008a. Late Quaternary vegetation and climate from temperate zone of the Kumaun Himalaya, India (with remarks on neotectonic disturbance). *Acta Palaeobotanica* 48 : 325-333.

Gupta, A. 2008b. Organic Matter and Climate since Middle Holocene from Himalaya (Talli Tal) – Geochemical Analysis. *International Symp. Plant Biol. and Envinorment Changing Scenario Abstr.* : 322.

Gupta, A. 2008c. Organic-Matter and Climate since Early Holocene, from Kumaun Himalaya – Geochemical Analysis. *Conf. Plant Life Through Ages Abstr.* : 58.

Gupta, A. 2009. Study of Holocene Floral and Faunal Remains revealed concealed Neotectonic Disturbance (Saria Tal, Kumaun Himalaya, India). *Acta Geologica Sinica (English Edition)*, 83 : (in press).

Gupta, H.P. 1977. Pollen analytical reconnaissance of Post glacial deposits from subtropical zone in Naini Tal District, Kumaun Himalaya. *Palaeobotanist* 24 : 215-244.

Gupta, H.P. & Khandelwal, A. 1982. Late Holocene Palynology from the lake in Sat Tal Valley, district Naini Tal, Kumaun Himalaya, U.P. *Geophytology* 12 : 313-321.

Gupta, R.K. 1968. *Flora Nainitalensis – a hand book of the Flowering plants of Nainital.* Navayug traders, New Delhi.

Heiri, O., Lotter, A.F. & Lemcke, G. 2001. Loss on ignition as a method for estimating organic and carbonate content in sediments : reproducibility and comparability of results. *Journal of Paleolimnology* 25 : 101-110.

Hukku B.M., Srivastava A.K. & Jaitle, G.N. 1974. Evolution of Lakes Around Naini Tal and the Problem of Hillside Instability. *Him. Geol.* 4: 516-531.

Jain, S.K. 1956. On a Botanical trip to Nainital. *Indian Forester* 82 : 22-38.

Kotlia, B.S., Bhalla, M.S., Sharma, C., Rajagopalan, G., Ramesh, R. Chauhan, M.S., Mathur, P.D., Bhandari, S. & Chacko, S.T. 1997. Palaeoclimatic conditions in the upper Pleistocene and Holocene Bhimtal-Naukuchiatal lake basin in south-central Kumaun, North India. *Palaeogeogr. Palaeoclimat. Palaeoecol.* 130 : 307-322.

Middlemiss, C.S. 1890. Geological sketch of Naini Tal; with some remarks on the natural conditions governing mountain slopes. *Rec. Geol. Surv. India* 23 : 213-234.

Mittre, V., Gupta, H.P. & Robert, R. 1967. Studies of the Late Quaternary vegetational history of Kumaun Himalaya. *Current Science* 36 : 539-540.

Mooney, S. 1997. A fine resolution paleoclimatic reconstruction of last 2000 yrs. From lake Keilambete, southeastern Australia. *The Holocene* 7 : 139-149.

Osmaston, A.E. 1927. *A Forest Flora of Kumaun.* Allahabad.

Pal, D. & Merh, S.S. 1974. Stratigraphy and structure of the Naini Tal area in Kumaun Himalaya. *Him. Geol.* 4 : 547-562.

Pal, D. & Merh, S.S. 1975. Geomorphic Evolution of the Naini Tal area, Kumaun Himalaya. *Him. Geol.* 5 : 327-338.

Pande, I.C. 1974. Tectonic Interpretation of the Geology of Naini Tal Area. *Him. Geol.* 4 : 532-546.

Pandey, U., Saxena, A.K., Pande, H.K. & Singh, J.S. 1984. Problems of forests and forestry in Kumaun Himalaya. *In* : The vegetational wealth of Himalaya (Ed. G.S. Paliwal) Puja Publisher, Delhi : 68-90.

Raina, B.N. & Dungrakoti, B.D. 1975. Geology of the area between Naini Tal and Champawat, Kumaun Himalaya, Uttar Pradesh. *Him. Geol.* 5: 1-28.

Tiwari, A.P. & Mehdi, S.H. 1964. Geology of Naini Tal – Almora Himalaya, U.P. Guide Ex. No. A-4, Inernt. Geol. Congr. XXII, Session, New Delhi : 1-19.

Valdiya, K.S. 1980. *Geology of Kumaun Lesser Himalaya.* WIHG Publ., The Himachal Times Press, Dehradun.

Valdiya, K.S. 1988. *Geology and Natural Environment of Nainital Hills, Kumaun Himalaya.* Gyanodaya Prakashan, Naini Tal.

Valdiya, K.S., Joshi, D.D., Sanwal, R. & Tandon, S.K. 1984. Geomorphologic development across the active main boundary thrust, an example from the Naini Tal Hills in Kumaun Himalaya. *J. Geol. Soc. India* 25 : 761-774.

❑❑❑

Chapter 6

Kaleidoscopic Behaviour of *Spirulina* Under Different Culture Conditions

SUDHIR D. GHATNEKAR AND SANTOSH M. SHARMA

INTRODUCTION

Dedicated research specific to study latent beneficial properties of *Spirulina,* a blue–green micro alga has revealed that it possesses abundant and varied nutrition in a small biomass. It has high protein content [70%], ideal combination of vitamins and minerals and important pigments like beta-carotene, phycocyanin etc. *Spirulina* is the only food source, except mother's milk, which contains a substantial amount of essential fatty acid viz., ã - linolenic acid [GLA]. This component makes *Spirulina* an ideal food supplement with therapeutic value and hence is considered as the coveted probiotic nutrient of the 21st century [Ghatnekar *et al* 2009]. *Spirulina* has been commercially most valued micro alga because of its pharmaceutical formulations, fine chemicals and easy cultivation. It also has the ability of bio-management of effluents from diverse industries if proper technology is employed.

However, *Spirulina* exhibits kaleidoscopic behaviour in its morphological and physiological parameters when subjected of different culture conditions. The present paper highlights the intensive work carried out for the characterization of *Spirulina* using morpho- physiological approaches. It also presents the developments being adopted by the biotechnology industry for maximizing its commercial exploitation.

SPIRULINA - TAXONOMIC CONSIDERATION

Spirulina is free-floating, filamentous phytoplankton. Due to the presence of green pigment chlorophyll a, like higher plants, botanists classify *Spirulina spp.* as a blue – green, photoautotrophic, non-heterocystous micro alga belonging to family Oscillatoriaceae under class Cyanophyceae, order Nostocales.

Biotechnology Resource Centre, G/1, Adinath, Shaikh Misry Road, Wadala (E), Mumbai–400 037.

However, due to their prokaryotic structure, they have been officially accepted by *Bergey's Manual of Systematic Bacteriology* as an oxygenic photosynthetic bacteria placed in group Cyanobacteria, Order Oscillatoriales [Castenholz 1989].

Arthrospira-Spirulina?

Spirulina refers to the popular trade name given to the dried biomass of the cyanobacterium *Arthrospira platensis and A. maxima* [Belay 1997 b, Webliography 1]. In the early 1930s, Geitler's taxonomic treatise for the classification of Cyanobacteria proposed the combination of all strains with helical trichomes into a single genus viz., *Spirulina*. Hence the species of the genus *Arthrospira* which were earlier placed separately [Gomont 1892] were again grouped in the genus *Spirulina*.

In recent times the view that *Arthrospira* and *Spirulina* are separate genera has been shared by several authors [Desikachary, 1959; Guglielmi *et al*., 1993]. The commercially important cultivated species viz., *Spirulina platensis* and *S. maxima* are currently included in genus *Arthrospira* Stizenberger 1852 [Castenholz 1989, Guglielmi *et al*., 1993] and there is an agreement that genus *Spirulina* is in fact different and also phylogenetically distant from genus *Arthrospira*.

Both the genera share a common characteristic of helical, non-heterocystous and cylindrical trichome and are placed in *the order Oscillatoriales*. However, the two genera can be delineated on the basis of morphological characteristics some of which are enumerated below.

1. The pitch of trichome helix from the transverse axis is <45 in *Spirulina* while it is >45 in *Arthrospira* [Castenholz 1989].
2. Genus *Spirulina* includes the individuals which seem unicellular since the cross walls of the cells of trichome are not visible while, *Arthrospira* filaments have cells separated by clearly visible septa [Desikachary 1959, Castenholz 1989].
3. *Spirulina sp.* shows 4–5 rows of pores arranged in hemi-circular arrays next to cross wall whereas, *Arthrospira sp.* exhibits presence of only one circular row of junctional pores [Gugliemi and Cohen-Bazire, 1982, Castenholz 1989, Gugliemi *et al*. 1993].

PHENOTYPIC CHARACTERS OF *SPIRULINA [ARTHROSPIRA]*

Morphology

The body of *Spirulina [Arthrospira]* is characterized by cylindrical trichome with multicellular construction in an open left-hand helix, due to which it derives its name. It is non - motile, but occasionally glides with a rotating motion. Each trichome may be slightly constricted at the cross walls. The trichome width varies from 3-12 μm in a variety of forms [Castenholz 1989] however the width can occasionally reach 16 μ [Belay 1997b]. The trichome end contains only little or no attenuation. The helix diameter ranges from 35-60 μ [Ciferri, 1983]. The cells in *Spirulina [Arthrospira] platensis* are nearly as long as broad, or shorter than broad to quadrate. Each cell is 2-6 μ long with granulated cross walls and end cells are broadly rounded [Desikachary 1959].

Ultrastructure

Under electron microscope *Spirulina* exhibits a typical prokaryotic structure devoid of membrane bound nucleus and cell organelles. However, thylakoids are observed as white strips distinct from well-evident mesosomes (Allen 1972, Tomaselli 1976]. Phycobilisomes [light harvesting complexes] are seen as black spots alternating with thylakoids. Presence of numerous patches of aggregated gas vesicles, polyglucan granules, cylindrical bodies, carboxysomes, and have also been detected.

ECOLOGY OF *SPIRULINA*

Spirulina has worldwide distribution in fresh water, marine and brackish waters and saline lake environments of tropical and semitropical regions [Castenholz 1989]. It grows luxuriantly in hot and alkaline [pH 8 -11] medium with high concentration of carbonates and bicarbonates at temperature of 32°C to 35°C.

CULTIVATION OF *SPIRULINA*

Spirulina has been cultured artificially by providing optimum requirements of nutrients, light, temperature, aeration etc. It is usually cultivated in a media composed of inorganic chemicals like sodium bicarbonate, a source of nitrogen, phosphorus, iron and trace metals.

Morpho-physiological Variations in *Spirulina* under Different Culture Conditions

Morphological characteristics of *Spirulina* species are often related to product quality e.g. the helicity has been used as a way to select high-quality strains [Belay 1997a]. In addition, the length of the spiral filament is very important for harvesting efficiency. Morphological changes in *S. platensis*, such as helix pitch, spiral width, or orientation, have been reported under various laboratory conditions [Jeeji Bai, and Seshadri, 1980, Jeeji Bai, 1985., Lewin, 1980].

The culture conditions such as light, temperature and composition of nutrients in the media significantly affect its morphology. They also interfere with the physiological features of this alga which might affect the nutraceutical value of *Spirulina* [Tamhane and Ghatnekar, 2004].

Effect of Physical Environment

Light : *Spirulina* is an obligate photoautotroph, hence depends on light as its main energy source. Varying intensity and duration of light definitely affects its phenotypic features as well as growth rate. *Spirulina* grown in bright sunshine areas are of best qualities probably due to higher beta-carotene production. Hongyan *et al* 2005 suggested that the breakage of filaments caused by solar Ultraviolet radiations [UVR] in outdoor cultures negatively influences the quality and yield of the harvested biomass.

Vonshak and Guy 1988 were first to describe the phenomenon of photo-inhibition in outdoor *Spirulina* cultures. They observed that shading could increase the photosynthetic activity and productivity. Studies conducted by Manoj Kumar *et al.* 2008 suggest that continuous illumination on the culture leads to lesser productivity, due to suppression of synthesis of chlorophyll-a , in relation to alternate light and dark period conditions.

They also observed its exceptional growth with high pigment content in natural sunlight at 29-36°C than in artificial light in laboratory conditions.

Light also induces morphological changes on the helicity of the *Spirulina* filaments. It seems that laboratory conditions, without any impact of UVR, favor the development of straight forms [Jeeji Bai, and Seshadri, 1980, Lewin, 1980, Jeeji Bai, 1985]. It was also observed that *S. platensis* strains with a tight helical structure could tolerate higher light intensity than strains with loose spirals [Jeeji Bai, and Seshadri, 1980] and that loose straight filaments could be transformed to tight coiled shapes when the cells were shifted to high light conditions [Fox, R. D. 1996]. These changes are perhaps associated with the protective strategy to counteract solar UVR or high Photosynthetically active radiation [PAR] levels by increasing self-shading [Hongyan *et al* 2005].

A fluorescent substance identified as biopterin-á-glucoside was isolated from the *A. platensis* cells. This substance prevented decolorization of the photosynthetic pigments, chlorophyll-a, phycocyanin and carotenoids in photosynthetic vesicles [Yukinori N. *et al* 1999].

Temperature

Temperature affects the metabolic activities and the growth rate, which makes it one of the most important limiting factors in *Spirulina* production. The usual temperature range for laboratory cultivation is 35-38^0 C. However certain thermophilic strains can resist up to 42^0 C. When certain strains have slow growth rate at 25^0 C, some show a tendency to turn yellow above 40^0 C and reduce yield [Tadros and MacElroy 1988].

Polyglucan granules were most prominent at low temperatures (15 to 17°C), decreasing in concentration at higher temperatures. At low temperatures, when the demand for amino acids is limited by the reduced growth rate, cyanophycin granules are the most abundant organelles, occupying up to 18% of the cell volume. On increasing the temperature, their number progressively decreases, and at 25-30°C, they are practically undetectable. However the relative amount of these granules also varies with the medium composition, and the cell age [Eykelenburg, 1980].

Cohen *et al.* 1987 reported an increase in the fatty acid content of the tested strains of *Spirulina* when the cultivation temperature was increased. The fatty acid content reached the highest value for most strains at 30-35^{0}C, and the fatty acids became more saturated. The alternate light and dark regime with high day and low night temperature required by *Spirulina* for optimum growth suggests its preference for thermoperiodicity [Tamhane and Ghatnekar 2004].

Spirulina has frequently been observed with a linear morphology in laboratory and industrial conditions. Apiradee *et al.*, 2007 observed presence of the linear and spiral forms of *S. platensis* C1 strain simultaneously. Their study concluded that environmental stress in the culture conditions and cell shape changing mechanism possibly induced morphological transformation in the selected strain.

The practical implication of these studies lies in the fact that, chemical compositions of *Spirulina* can be altered as desired by changing the physical conditions.

Media State and Helicity of Trichome

The helical shape of the trichome is characteristic of the genus but the helical parameters (i.e., pitch length and helix dimensions) vary with the species, and even within the same species (Marty and Busson, 1970]. The helical shape is maintained only in liquid media, and in solid media the filaments become true spirals [Eykelenburg and Fuchs, 1980]. The transition from helix to spiral is probably related to the necessity of reducing, in solid media, the surface area exposed to air. It is possible that these transitions are caused by hydration or dehydration of the oligopeptides in the peptidoglycan layer of the cell wall, resulting in changes in the rigidity of the cell [Eykelenburg, 1980b].

These observations reveal the kaleidoscopic behaviour of *Spirulina* with respect to its morpho-physiological parameters when subjected to varied culture conditions. This knowledge shall enable the cultivators to optimize the growth conditions for maintaining its natural morphology and physiological features and thereby enhance its productivity and nutrient value.

Environment Friendly Cultivation of *Spirulina*

In recent times, a need for using biological, environment friendly ingredients in the culture media for *Spirulina* cultivation has been emphasized. Tamhane and Ghatnekar 2004 studied the subsequent effect of chemicals vis à vis natural ingredients present in the culture media of *Spirulina.* They conducted series of experiments where simple biological ingredients such as potato extract, tea extract, cow's urine and coconut water were used as nutrient supplement for enhancing growth of *Spirulina.* They concluded that all four natural ingredients are suitable for enhancing growth and nutritional quality of *Spirulina.*

Biotechnology Resource Centre [BRC], under the leadership of Dr. Sudhir Ghatnekar, has developed standard culture medium "BRC-Spiro" using biological, environment friendly, bio-safe and bio-ethical ingredients, for the commercial cultivation of *Spirulina* [Plate. 1]. The use of this medium enhances the growth rate by 40%, than *Spirulina* grown in culture media using standard inorganic chemical ingredients. Also the time for cultivation is reduced by at least 27% [Tamhane, *et al*. 2005].

The nutrient composition of *S. platensis* cultured in BRC-Spiro medium was compared to *Spirulina* cultured in standard inorganic chemical medium [SCM] viz, Zarouk's medium. *Spirulina* cultured in BRC-Spiro media exhibited 29 % higher protein content than *Spirulina* cultured in SCM. Analysis of overall composition of amino acids, other bio- active substances and vitamins revealed similar trend [Table 1].

Sharma *et al* 2008, conducted trials with *Spirulina* meal cultured at BRC, by feeding it to Tilapia and Indian Major Carps in monoculture and polyculture systems. The findings from these trials indicate that *Spirulina* feed could replace many of the conventional expensive ingredients in the standard fish feed.

BRC conducted series of trials to evaluate the nutritional effect of organically cultured probiotic *S. platensis* by adding it to local food items like idli, dhokla and beverages like sherbets. Inclusion of *Spirulina* in the daily diet of children of age 3-10 years for six months [Plate 2], exhibited a significant increase in optimum weight and height of these children, in relation to the control group [Ghatnekar *et. al.* 2009].

TABLE 1

Nutritional profile of *Spirulina platensis* cultured in BRC – Spiro medium in relation to Standard inorganic chemical medium [SCM] [Tamhane and Ghatnekar 2004]

Nutrients	Nutritional profile of *S. platensis* cultured in selected media	
	BRC – Spiro medium	SCM
Overall composition [% dry matter]		
Crude Protein [N x 6.25]	71 - 75	55 – 58
Lipids 4 - 5	2 – 3	
Carbohydrates	14 - 15	10 – 11
Crude fibre	3 – 5	1 - 4
Ash 10 – 12	6 - 8	
Moisture 8 – 9	8 – 9	
Analysis of certain bio active compounds [g 100g $^{-1}$ dry matter]		
Total Nitrogen	12.7	10.0
Non – protein nitrogen	1.9	1.5
Available lysine [g/100g protein]	3.13	2.96
RNA 3.10	2.90	
DNA 1.40	1.00	
Calcium 0.82	0.75	
Phosphorus	1.59	1.42
Sodium 0.48	0.45	
Magnesium	0.97	0.90
Iron 0.17	0.12	
Potassium	1.89	1.42
Vitamins [mg Kg $^{-1}$ dry matter]		
Thiamine 30.9	27.8	
Riboflavin	37.8	33.4
Pyridoxyl - HCl	1.6	1.3
Cobalamine	2.8	2.4
Biotin0.01	0.06	
Beta – carotene	540	500

Plate 1 : Organically grown *Spirulina* using BRC - Spiro media

SPIRULINA – WASTE WATER TREATMENT

The most important aspect of *Spirulina* production is its ability to grow in impure water. This ability can be tapped for the bioremediation of domestic sewage and liquid effluents of various industries, which are loaded with large dose of complex nutrients necessary for its growth. BRC has devised a technology to cultivate *Spirulina sp.* in the liquid effluents from diverse industries. Thus, the wastes of the industries can be polished along with production of nutritional food supplement. [Ghatnekar and Kavian, 1992].

TABLE 2

Physical properties of *S. platensis* biomass harvested from different types of culture media

Observational Parameters	CONTROL [C 1] *Spirulina* cultured in traditional inorganic media	CONTROL [C 2] *Spirulina* cultured in BRC – Spiro media	TREATED [T] Spirulina cultured in effluents [50%]
Physical properties of harvested biomass			
Colour	Dark blue - green	Dark blue - green	Light green
Odour	Mild resembling sea weed	Mild resembling sea weed	Mild resembling sea weed
Taste	None	None	None
Consistency	Slimy when wet	Slimy when wet	Slimy when wet
pH	8	6.7	7.2
Fresh weight [gL^{-1}]	12.13	15.7	11.68
Insect fragments [10 g^{-1}]	<30	<30	<30

Plate 2 : Feeding *Spirulina* fortified idlis to children

Plate 3 : Experimental tanks at BRC used for biomanagement of gelatine industry effluents using combination of *Spirulina platensis, Aspergillus flavus* and *A. Niger*

Waste effluent from a fertilizer company in Nigeria was used for cultivation of *Chlorella* and *Spirulina.* Their study concluded that the non-sewage effluent could be used for production of microalgal biomass and value added biochemicals [Anaga and Abu, 1996].

Ghatnekar and his associate Sharma, 2008 attempted biomanagement of liquid effluents discharged after the secondary treatment from gelatine manufacturing industry using *S. platensis,* in combination with *Aspergillus flavus* and *A. niger* [Plates 3, 4]. In the above study, the liquid effluents were mixed in different concentrations with BRC-Spiro media and used for culturing *S. platensis*. Morphological features of *S. platensis*

Plate 4 : *Spirulina* as a test system in Effluents Treatment Plant

cultured in gelatine industry effluents were compared with *S. platensis* cultured in inorganic media and BRC – Spiro media [Table 2, 3]. BRC is now testing the efficacy of *Spirulina* cultured in effluents as a food probiotic in different test systems of mice and Tilapias.

TABLE 3

Morphological variations in *S. platensis* cultured in different types of media

Observational Parameters	CONTROL [C 1] *Spirulina* cultured in traditional inorganic media	CONTROL [C 2] *Spirulina* cultured in BRC – Spiro media	TREATED [T] Spirulina cultured in effluents [50%]
Morphological Analysis			
Average number of filaments [dilution 10^2]	140 ± 5.22	156.75 ± 8.44	74.25 ± 4.131
Average length of coiled trichome (μ)	127.88 ± 10.86	173.25 ± 7.84	108.75 ± 8.40
Helix diameter (μ)	28.13 ± 2.58	37.5 ±1.76	13.13 ± 4.30

[Value at 40 X magnification ± Standard error of mean of 10 samples]

Chang, 1999 studied the possibility of using nitrifying bacteria for the fulfillment of nitrogen fertilizer in *Spirulina* mass culture.

Ghatnekar *et al* 2009 suggested the possibility of integrating conventional agriculture in rural areas with *Spirulina* farming for multiple benefits for the villagers. For making *Spirulina* cultivation cost effective, organic manure from bio-gas plant and vermi-compost plant will be used with state-of-the-art technology instead of using inorganic chemicals. In fine, besides being an inexpensive agent for reclaiming the sewage, the cultivation of algae in sewage offers an economic potential of producing low cost algal protein. However, these effluents have to be carefully handled during *Spirulina* cultivation as they can also harbor pathogens.

CONCLUSIONS

Since early 1960s, the mass production of *Spirulina* has attracted the interest of the researchers due to its tremendous application in food, feed, pharmaceutical and cosmetic industry. Their major concern has been formulating innovative methods for increasing the yield while lowering the cost of its production. The study of factors affecting the kaleidoscopic behaviour of morpho-physiological features of *Spirulina* has been of tremendous advantage to the biotechnologists in deciding the optimum conditions required for enhancing its growth and quality. Such studies similar to those carried out by BRC, to culture *Spirulina* in locally available raw material and waste water has further enhanced the feasibility of its commercial production.

REFERENCES

Abo-Shady, A. M., Abou – El – Souod, S. M., El. Bakaem, A., El – Shanshoury, R. and Mahmoud, Y.A.G. 1992. Protoplast from the cyanobacterium *Spirulina platensis*. World *J. Microbiol. And Biotech*, 8: 385 – 386.

Apiradee H., Matura S., Peerada P., Sritana T., Boosya B., Supapon C., and Morakot T., 2007. Revealing differentially expressed proteins in two morphological forms of *Spirulina platensis* by proteomic analysis, *Molecular Biotechnology*, Vol. 36, No. 2 : 123-130.

Allen, M. M. 1972. Mesosomes in blue-green algae. *Arch. Mikrobiol.* 84:199-206.

Anaga, A., and Abu, G. O. 1996. A laboratory scale cultivation of Chlorella and Spirulina using waste effluent from a fertilizer company in Nigeria, *Biores. Technol.*, 58: 93 – 95.

Belay, A. 1997a. Mass culture of Spirulina outdoors—the Earthrise farms experience, In A. Vonshak (ed.), Spirulina platensis (Arthrospira): physiology, cell biology and biotechnology. Taylor & Francis Publishers, London, United Kingdom. 131-158.

Belay, A. 1997 b. *Spirulina [Arthrospira]* : Production and Quality assurance, M. E. Gershwin and Amha Belay[eds] *Spirulina* in human nutrition and health. CRC press.

Castenholz, R. W. 1989. Subsection III, Order *Oscillatoriales*, In Stanley, J. T., Bryant, M. P., Pfenning, N. and Holt, J. G. (Eds), Bergey's Manual of Systematic Bacteriology, Vol 3, Baltimore: William and Wilkins : 1771.

Chang, 1999. The possibility of nitrifying bacteria inoculation in *Spirulina* mass culture, Jiangsu *J. Agril. Sci.*, 15: 191 - 192

Ciferri. O., 1983. *Spirulina*, The edible Microorganism. *Micriobiol. Rev. vol* 47 : 558 – 559.

Cohen, Z., Vonshak, A., and Richmond, A. 1987. Fatty acid composition in different *Spirulina* strains under various environmental conditions, *Phytochemistry*, 26: 2255.

Desikacharya, T. V. 1959. Cyanophyta, ICAR monograph on algae, New Delhi, India: 187.

Eykelenburg, C. V. 1980a. Ecophysiological studies on *Spirulina platensis*. Effect of temperature, light intensity and nitrate concentration on growth and ultrastructure. Antonie van Leeuwenhoek *J. Microbiol. Serol.* 46: 113- 127.

Eykelenburg, C V. 1980b. Some theoretical considerations on the in vitro shape of the cross-walls in *Spirulina* spp. *J. Theor. Biol.* 82: 271-282.

Eykelenburg. C. V., and Fuchs, A. 1980. Rapid reversible macromorphological changes in *Spirulina platensis., Naturwissenschaften,* 67: 200-201.

Fox, R. D. 1996. *Spirulina.* Production and potential. Edisud, Aixen-Provence, France.

Ghatnekar, Sudhir D. and Kavian, Mahavash F. 1992. New Spins-offs of Biotech after the Rio-Earth Summit, Proceedings of inaugural session of *Biotech India* '92. Oct.28-31.

Ghatnekar, Sudhir D., Tamhane, Bharati M., Sawant, Sudha A and Sharma, Santosh M., 2008, Biomanagement of liquid effluents from gelatine manufacturing industry using combination of *Spirulina platensis, Aspergillus flavus* and *Aspergillus niger*. Paper presented at the XXXI All India Botanical Conference and International Symposium on Plant Biology and Environment: Changing Scenario, Department of Botany, University of Allahabad, December 17 – 19, 2008.

Ghatnekar Sudhir D., Sharma Santosh M. Sawant Sudha A. 2009. Inclusion of Organically grown *Spirulina platensis* as probiotic supplement in idli, Dhokla and beverage in the daily diet of children. Paper presented at the National conference on Biofunctions, Biodiversity and Plant Resource utilization. D.D.U. Gorakhpur University, January. 109 – 110.

Ghatnekar, Sudhir, D., Sharma, Santosh, M., Ghatnekar, Sumukh, S., Karnik, Deepika, A. 2009. *Spirulina* [Blue – green micro alga] – A probiotic nutraceutical food supplement for 21st century. *Green Farming* vol. 2 [5]: 335 – 338.

Gomont M., 1892. Monographie des Oscillatoriees, - Annales des Sciences Naturelles (Botanique). Serie &, Tome XVI: 91 – 264

Guglielmi, G. and Cohen-Bazire, G., 1982. Structure et distribution des pores et des perforations de l' envelope de peptidoglycane chez quelques cyanobacteries, - *Protistologica,* XVIII: 151–165.

Guglielmi G., Rippka R., N. Tandeau de Marsac. 1993. Main properties that justify the different taxonomic position of *Spirulina spp.* and *Arthrospira spp* among Cyanobacteria. Bulletin de I'Institut Oceanographique, Monaco, n^{0} special 12. 13 – 23.

Hongyan Wu, Kunshan Gao, Virginia E. Villafañe, Teruo Watanabe, and E. Walter Helbling., 2005. Effects of Solar UV Radiation on Morphology and Photosynthesis of Filamentous Cyanobacterium Arthrospira platensis. *Appl Environ. Microbiol.* 71(9): 5004–5013.

Hiroyuki, K., Qin, S., Thankappan, A. K., Yoshikazu K., Shin – ichi, Y., 1998. Transposable genetic elements in *Spirulina* and potential application s for genetic engineering, *Chin. J. Oceanol. Limnol.* Vol. 16, Suppl : 30 – 39.

Jeeji Bai, N. 1985. Competitive exclusion or morphological transformation? A case study with Spirulina fusiformis, Arch. Hydrobiol., Suppl. 71, Algol. Stud., 38 / 39: 191

Jeeji Bai, N. and Seshadri, C. V. 1980. On coiling and uncoiling of trichomes in the genus *Spirulina, Arch. Hydrobiol.,* Suppl. 60, *Algol. Stud.* 26: 32.

Lewin, R. A. 1980. Uncoiled variants of Spirulina platensis (Cyanophyceae: Oscillatoriaceae). *Arch. Hydrobiol.* (Suppl. 60 Algol. Stud.) 26: 48-52.

Manoj Kumar, Rawat, S., and Singh G. P., 2008. Impact of different culture conditions on growth and pigment contents of Spirulina platensis [Jal – Mahal Isolate] J. *Indian Bot. Soc.,* Vol. 87 [3 & 4]: 267 - 271

Marty, F., and F. Busson. 1970. Donnees cytologique, et systematiques sur *Spirulina platensis*

(Gom.) Geitler et *Spirulina* Geitleri J. de Toni (Cyanophyceae-Oscillatoriaceae). C. R. Acad. Sci. Ser. D 270: 786-789.

Sharma, Santosh, M., Tamhane, Bharati, M., Sawant, Sudha, A., and Ghatnekar, Sudhir, D., 2008. Feeding Spirulina to Mozambiq Tilapia and Indian Major Carps. JIBS 87 [3&4]. 248 – 251.

Tamhane Bharati M., Ghatnekar Sudhir D. 2004. Effect of chemicals vis-à-vis Natural ingredients in the Culture Media of *Spirulina Platensis*. Proceedings of National Seminar on new Horizons in Environmental Science and Engineering in India. University of Bangalore. 17 –19.

Tamhane Bharati M., Ghatnekar Sumukh S., Ghatnekar Sudhir D. 2005. *Spirulina*: Non – conventional Food. Financing Agriculture July – September : 44 – 47.

Tadros, M. G. and MacElroy, R. D. 1988. Characterization of Spirulina biomass for CELSS diet potential. NASA Cooperative Agreement CELSS program [Alabama A and M Univ.]: 53.

Tomaselli, F., L., M. C. Margheri, and E. Pelosi. 1976. Die Ultrastruktur von Spirulina in Vergleich zu Oscillatoria. Zentralbl. Bakteriol. Parasitenkd. Infektionskr. Hyg. Abt. 2 131:592-601.

Vonshak, A., 1997. *Spirulina*: Growth, Physiology and Biochemistry, In A. Vonshak (ed.), *Spirulina platensis* (*Arthrospira*): physiology, cell biology and biotechnology. Taylor & Francis Publishers, London, United Kingdom : 67 - 77.

Vonshak, A., and Guy, R., 1988. Photoinhibition as a limiting factor in outdoor cultivation of *Spirulina platensis*. In Stadler, T., Mollion, J., Verdus, M. C., Karamdas, Y., Morvan, H., and Christiaen, D., [Eds] *Algal Biotechnology*: 365 – 370, London: Elsevier Applied Science.

Yukinori N., Asako I., Ayako M., Daisuke H., Ken-ichi Y., Tomohisa M., Takeshi W., Yoh K., Misao H., Hiroyuki N., Mitsuo S., and Yuji I., 1999. Isolation of Biopterin-á-glucoside from Spirulina (*Arthrospira*) platensis and Its Physiologic Function. *Marine Biotechnology*., Vol 1(2): 207-210.

❑❑❑

Chapter 7

Plant Extracts–A Tool for Genotoxic Damage Repair

NANDJEE KUMAR, V.S. SINHA[1] AND V.K. PRABHAT

INTRODUCTION

Pesticides are in wide range of application in order to control weeds, insects and fungi and to enhance the agricultural yields. Several million tons of organic and inorganic chemicals with antimicrobial and insecticidal properties are added annually to soil and their environment (Jain and Sarbhoy, 1988).Grant (1978) pointed out that plant system appears to be an excellent indicator of mutagenic / cytotoxic / genotoxic effects of chemical mutagens, therefore the plant system must be accepted as a first tier assay for detection of the possible genetic damage resulting due to environmental chemicals. Hence, environmental biologists are showing great concern over the indiscriminate use of synthetic chemicals. Some of them not only upset the ecosystem, rather produce undesirable changes both physiologically as well as genetically in crop plants.

The mutagenic and/or carcinogenic potential to the non-target organism have been demonstrated by several workers (Grant, 1978; Badar, 1983; Ahmad and Yashmin, 1992). The mitodepressive and chromotoxic activities of pesticides in several crop plants have also been reported by Badar (1983).In recent years allelopathy has acquired a key role to study the mutagenic / genotoxic / chromotoxic / cytotoxic effects on living organism and plant extracts have been utilized as an alternative to pesticides so called biopesticides. Various workers reported remarkable allelopathic effect on cytotoxicity / chromotoxicity in Neem (Awasthy *et al.* 1995; Sinha and Kumar, 2005), mustard (Kumar and Kohli 1987).,*Lantana*, (Sinha, Kumar and Mohanka 2005), *Chrysanthemum* (Sinha, 2003).

The present paper deals with the antimutagenic potential of the plant extracts viz

Department of Botany, Magadh University, Bodhgaya-824 234 (India)
[1]Department of Botany, Tata College, Chaibasa-833 202 (India)

TABLE 1

Effect of chemical pesticides on MI and % chromosomal abnormalities on *Vicia faba* L.

Treatment	Concent-ration (%)	No. of dividing cells	Abnormal cells(%)	Fragments (%)	Stickiness (%)	Disturbed metaphase/ anaphase(%)	Bridges (%)	Laggards (%)	Micronu-clei (%)	Mitotic Index (%)
Control	-	445	0.31	0.16	-	-	-	0.15	-	25.65 ± 0.12
R	0.12	356	9.59	2.33	3.07	1.88	1.23	0.88	0.20	14.80± 0.87
E		378	11.85	1.22	2.25	1.67	4.56	1.70	0.45	12.60± 1.76
D		434	11.83	1.66	1.65	1.78	2.98	2.75	1.01	11.90± 1.90
R	0.25	254	20.23	3.50	4.65	1.90	4.78	3.75	1.65	12.12± 1.22
E		343	26.00	4.45	6.67	1.89	5.77	5.45	1.77	10.11± 0.94
D		365	21.55	2.26	4.45	2.20	6.66	4.88	2.10	09.67± 1.43
R	0.50	132	25.82	3.33	5.88	4.40	7.87	5.70	4.52	9.88 ± 1.54
E		234	30.17	4.42	7.85	3.86	6.90	4.88	2.26	7.88 ± 0.78
D		325	32.64	3.88	6.85	4.78	7.98	5.60	3.55	8.84 ± 1.77
R	1.00	167	58.97	6.76	12.65	8.55	12.69	12.45	5.87	6.78 ± 0.98
E		237	57.14	5.68	15.50	9.80	9.95	11.44	4.77	6.88 ± 1.14
D		316	63.18	6.77	16.90	6.44	14.54	10.65	7.88	8.87 ± 0.78
R	2.00	121	29.39	3.65	5.75	4.23	5.65	5.67	4.44	12.43 ±0.84
E		187	28.16	2.20	4.45	4.88	4.24	8.52	3.87	14.56 ± 0.78
D		239	21.30	1.65	5.99	2.58	2.98	3.66	4.44	11.82± 0.69

TABLE 2

Synergistic effect of chemical pesticides and plant extracts on MI and % chromosomal abnormalities in *Vicia faba* L.

Treatment	Concent-ration (%)	No. of dividing cells	Abnormal cells(%)	Fragments (%)	Stickiness (%)	Disturbed metaphase/ anaphase(%)	Bridges (%)	Laggards (%)	Micronu-clei (%)	Mitotic Index (%)
Control	-	450	0.18	0.08	-	-	-	0.10	-	27.87±1.97
A	0.12	423	17.67	2.01	3.54	2.67	3.45	4.55	1.45	39.98± 1.02
B		378	13.44	3.06	4.33	1.54	2.87	4.76	1.21	38.56 ± 0.88
C		344	15.39	1.55	3.10	1.44	3.00	4.87	1.43	40.87 ± 1.02
A	0.25	325	15.25	2.56	2.23	1.23	4.65	3.34	1.24	40.45 ± 1.01
B		239	14.43	2.67	1.88	1.54	3.44	3.14	1.76	43.87 ± 3 .09
C		323	13.64	1.98.	2.76	1.78	3.23	2.45	1.44	35.34 ± 3.65
A	0.50	366	10.35	1.89	2.00	1.02	2.99	1.34	1.11	45.65 ± 3.78
B		275	11.49	2.33	2.12	1.43	2.68	1.99	0.94	39.88 ± 1.20
C		180	08.77	1.66	1.88	1.24	2.76	0.89	0.34	41,20 ± 1.13
A	1.00	198	03.46	0.23	0.56	0.45	1.65	0.34	0.23	48.90 ± 2.43
B		328	03.27	0.10	0.87	0.65	1.33	0.20	0.12	46.80 ± 4.34
C		333	02.20	0.21	0.46	0.11	1.20	0.12	0.10	44.65 ± 5.25
A	2.00	447	07.88	0.99	1.21	1.30	1.30	1.76	1.32	33.79 ± 3.20
B		324	08.62	1.77	1.38	1.87	1.21	1.34	1.05	38.70 ± 2.89
C		269	09.11	1.54	1.48	1.56	1.27	1.96	1.30	29.90 ±3.32

Azadirachta indica A Juss(A), *Blumea lacera* (B) and *Chrysanthemum cinerarifolium* Vis (C) along with chemical pesticides on the meristematic root tip cells of *Vicia faba* L.

MATERIALS AND METHODS

Preparation of pesticides solution: The pesticide solutions of varying concentrations (0.12%, 0.25%, 0.50%, 0.75%, 1.00% and 2.00 %) were prepared by dissolving desired amount of Rogor, Endosulphan and Dichlovos in distilled water.

Preparation of plant extract: 500 gm of Leaves of *Azadirachta indica* , *Blumea lacera* and *Chrysanthemum cinerarifolium* were washed thoroughly and chopped into small pieces and grinded by using an electric mixer with 200 ml water. The entire contents were stirred for 3hrs in a magnetic stirrer and filtered to get mother solution with 100% concentration which was further diluted to different concentrations such as 0.12%, 0.25%, 0.50%, 1.00% and 2.00 % by adding desired amount of distilled water.

Squash Preparation: The seeds of *Vicia faba* L were treated with different concentrations (0.12%,0.25%, 0.50% , 1.00% and 2.00%) of three chemical pesticides viz Rogor (R) , Endosulphan(E) and Dichlorovos(D) and of a mixture of the chemical pesticides as well as plant extracts of *Azadirachta indica* , *Blumea lacera* and *Chrysanthemum cinerarifolium* separately for four hours. The treated seeds were allowed to germinate on moist petridishes and the root tips were squashed and stained in 2 % acetocarmine solution.

The genotoxic effects of chemical pesticides and different plant extracts were analyzed by calculating mitotic index (MI) and different types of percent chromosomal abnormalities at metaphase and anaphase stages of root tips by using the formulae mentioned as under;

MI = Number of dividing cells/Total number of cellsx100 and % Chromosomal abnormalities= Number of abnormal cells/Total number of dividing cellsx100

RESULTS AND DISCUSSION

In *Vicia faba* L, the number of somatic chromosomes was noted to be 12.The cytological analysis revealed that the chromosomal abnormalities were almost negligible in control plants; however the treated root tips showed considerable amount of chromosomal abnormalities. It was interesting to note that the chemical pesticides gradually reduced the mitotic index but induced gradually the chromosomal abnormalities significantly. The maximum abnormalities and minimum mitotic index were recorded in 1.00% concentration of the pesticides. Surprisingly the mixture of the chemical pesticides and the plant extracts gradually induced the mitotic index and gradually reduced the chromosomal abnormalities. Interestingly 1.00 % concentration of the mixture of the chemical pesticides and the plant extracts significantly reduced the chromosomal abnormalities, thus the plant extracts restored the normal chromosomal behavior in *Vicia faba* which indicates that the plant extracts of *Azadirachta indica* ,*Blumea lacera* and *Chrysanthemum cinerarifolium* have strong potential to repair the genotoxic damages.

The genotoxic effect has earlier been carried out in *Allim cepa* (Badar, 1983; Ahmad and Yasmin, 1992), Barley (Vaarama, 1947; Yakar, 1952) and wheat (Soliman and Al Najjar, 1980).They have pointed out that the chemical pesticides do cause chromosomal anomalies such as c-mitosis, fragments, stickiness, bridges, laggards and micronuclei.Kaur and Grover (1985) studied the genotoxic effect of organophosphate pesticides and reported the chromosomal abnormalities.

The different chromosomal abnormalities noted in the present investigation were fragments, stickiness, disturbed metaphase/anaphase, bridges, laggards and micronuclei. The synergistic effect of chemical pesticides and the plant extracts for the amplification of mitotic index may be helpful in inducing cell division in plants. The gradation of the synergistic effect was noted as A+D>A+E>A+R>B+D>B+E>B+R>C+D>C+E>C+R. The genotoxic agents such as chemical pesticides may cause metabolic imbalances which may interfere the structure and synthesis of nucleic acid. The genotoxicity of the pesticides may cause the deviation of normal metabolic activities from its original parental line.

Obviously, the plant extracts repair the damages caused by the chemical pesticides to a considerable extent and may be better substitutes of the chemical pesticides in view of reduced risk of genotoxic hazards to the agriculturally important crops.

REFERENCES

Ahmad, S. and Yasmin, R. 1992. Effects of methyl parathion and Trimiltox on the mitosis of *Allium cepa*. *Cytologia* 57: 155-160

Awasthy, K. S., Chaurasia, O. P. and Sinha, S. P. 1995. Genotoxic effects of crude extract of neem in bone marrow cells of mice. *Cytologia* 60: 189-193.

Badar, A. 1983. Cytogenetic activities of Triazma herbicide in root tips of *Allium cepa* and *Vicia faba. Mut Res* 117: 173-182.

Grant, W. F. 1978. Chromosome aberrations in plants as monitoring system Env Health. *Pers* 27: 289-293.

Jain, A. K., Ajay, K. J. and Sarbhoy, R. K. 1988. Cytogenetic studies on the effect of some chlorinated pesticides. *Cytologia* 53: 427-436

Kaur, P. and Grover, I.S. 1985. Cytological effect of some organophosphorus pesticides, *Cytologia* 50: 187-197.

Kumar, A. and Kohli, R. K. 1987. Autotoxicity of Reg weed *Parthenium. Weed* Sc 15: 629-632.

Sinha, V. S. 2003. Comparative studies of the hazardous effects of the chemical and biological pesticides on *Vicia faba* L Ph.D. thesis of Patna University, Patna.

Sinha, V. S. and Kumar, Nand Mohanka R 2005. Neem an alternative to synthetic pesticide International Symposium on Diet in causation and preventation of cancer and XXX Annual conference of Environmental Mutagen Society of India.

Siliman, A.L. and A.l. Najjar, N. R. 1980. Cytological effects of fungicides on wheat and two related species. *Cytologia* 45: 160-163

Vaarama, A. 1947. Experimental studies on the influence of DDT insecticides upon Plant 33: 191-219.

Yakar, N. 1952. Mitotic disturbances caused by Chloranil. *Amer J Bot.* 33:540.

Chapter 8

An Approach to Identify Different Varieties of *Mangifera indica* L. with the Help of Morphological Traits

BHAVNA SHARMA, SUSY ALBERT AND HARESH DHADUK

INTRODUCTION

Mango is commonly known as King of Fruits because of its excellent flavour, delicious taste, attractive appeareance, high nutritive value, richness in the varieties and popularity among masses. Its development and culture in the sub-continent were primarily contributed by the Mughal Emperors, particularly Akbar who planted Lakh Bagh. India is the largest producer (12000 million tons) followed by China (2142 million tons) and Mexico (FAO 2000). Mango, the most important fruit of India, is grown in an area of 1.23 million ha with an annual production of 10.99 million tonnes, which accounts for 57.18 per cent of the total world production (Negi, 2000). Due to long history of cultivation about a thousand cultivars of mango are known to exist in India.

Studies on the vegetative and floral development of mango are essential as the fruit growth rates increased with higher leaf: fruit ratio. Leaf: fruit ratio also affects the eating quality Levels of carbohydrates, especially starch, sucrose and fructose, were higher when assimilating supply increased (100 leaves per fruit). Citric and malic acids metabolisms also show influenced by the leaf: fruit ratio (Lechaudel, Joas & Jannoyer 2004). The initial fruit set is directly related to the proportion of perfect flowers. Proportion of perfect to staminate flower (sex ratio) is a variable component within panicles, trees and among cultivars. (James, Marray & Gaskins, 1961). The performance of varieties is found to vary under different climatic conditions (Singh, 1978).Variations appear to be existing in the number of perfect flowers, fruit set and extent of fruit drop in different varieties of mango (Anila and Radha, 2003). Some varieties, which are important either from commercial or from breeding point of view, were studied for their morphogenetic variations. These

Department of Botany, Faculty of Science, The M.S.University of Baroda, Vadodara - 390 002, Gujarat.

TABLE 1
Morphological features of leaf, inflorescence and fruit

Features	Varieties																
	Alphonso	Badshah pasand	Goto	Sindoria	Rucchado	Desi	Dudh pendo	Fasli	Jamadar	Cowasji	Kesar	Khodi	Totapuri	Pairi	Rajapuri	Sopari	Langd
Leaf																	
Pulvinus leaf base (cm)	1.42	1.6	1.38	1.38	1.7	1.38	1.46	0.78	0.98	1.96	1.5	1.2	1.58	1.44	1.44	1.04	1.66
Petiole (cm)	2.08	2.46	1.4	2.42	2.06	2.28	4.24	1.76	2.3	1.92	2.02	2.1	1.64	2.34	1.84	1.32	1.28
Length(cm)	21.52	16.78	24.05	20.1	17.4	19.31	39.97	16.78	16.5	26.78	30.8	23.7	23.04	24.9	18.68	17.38	18.72
Breadth(cm)	4.5	4.8	7.1	5.32	5.78	4.7	7.05	2.6	3.6	6.46	4.77	5.6	5.7	6.38	6.7	3.22	5.82
Inflorescence																	
Color	Green	Yellow	Reddish yellow	Yellow	Light green	Green	Red	Red	Yellow	Yellow	Reddish yellow	Yellow -ish green	Yellow	Reddish yellow	Green	Reddish yellow	Yellov
Length(cm)	12	20-22	20-25	20-22	29-30	28-30	30-32	17-20	38-40	13-15	33-35	28-30	20-22	38-40	22-25	20-22	20-25
Ratio (B:M)	40:60	60:40	5:95	5:95	90:10	20:80	60:40	60:40	40:60	20:80	80:20	50:50	50:50	70:30	80:20	85:15	90:10
Fruit																	
Color	Green with Red blush	Green	Green	Green	Green	Yellow -ish green	green	Green	Green	Green with Red blush	Green	Green	Yellow	Green with Red blush	Green	Green	Greer
Weight(g)	153.33	228.33	282.5	106.66	176.66	123.3	135	355	200	436.7	305	72.5	352.5	245	440	48.33	226.6
Length (cm)	10.1	11.53	14	10.75	13.75	9.4	9.6	14	11.5	15.7	1505	8.45	16.75	12.3	14.75	7	12.4
Diameter(cm)	19.55	21.6	22.6	15.2	18.5	17.2	18.8	23.75	21	25.33	22.5	14.3	22.45	23.16	27.75	13	20.75

B: Bisexual & M: Male flowers.

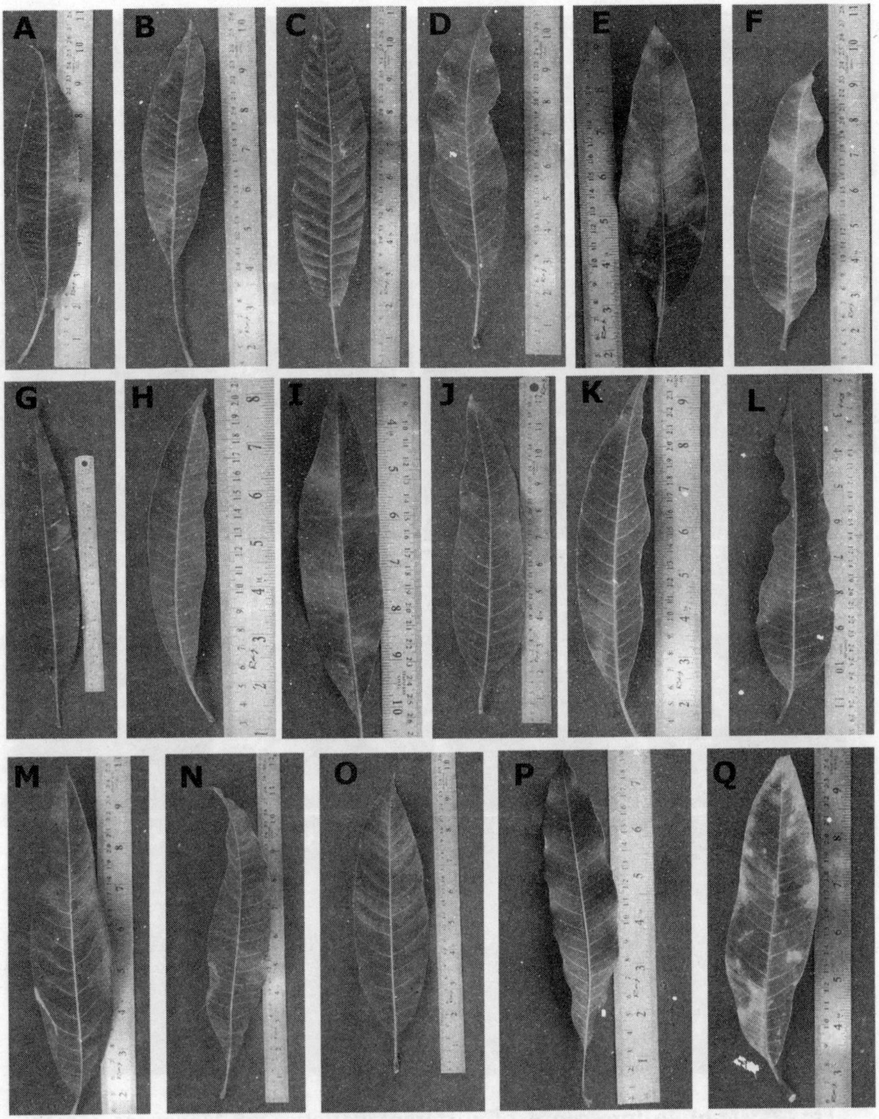

Fig. 1 : (A-Q) Leaf of different varieties of Mangifera indica.

A. Leaf of Alphonso with a prominent midrib, slightly wavy margin & acute to acuminate apex. B. Leaf of Badshashpasand with long petiole, wavy margin & acute apex. C. Leaf of Goto with the lateral veins deeply placed, wavy margin & acute apex. D. Leaf of Sindoria with a wavy margin & acute to acuminate apex. E. Leaf of Rucchado with a smooth margin and acute apex. F. Leaf of Desi with very short petiole, wavy margin & acute to acuminate apex. G. Leaf of Dudhpendo with a smooth margin & acute apex. H. Leaf of Fasli with a slightly wavy margin & acute apex. I. Leaf of Jamadar with a slightly wavy margin & acute apex. J. Leaf of Cowasji with a short petiole, a broad base, slightly wavy margin & acute to acuminate apex. K. Leaf of Kesar with wavy margin & acute to acuminate apex. L. Leaf of Khodi with a prominent midrib, very wavy margin & acute apex. M. Leaf of Totapuri with short petiole, slightly wavy margin & acute apex. N. Leaf of Pairi with a wavy margin & acute apex. O. Leaf of Rajapuri with slightly wavy margin & acute apex. P. Leaf of Sopari with a wavy margin & acute apex. Q. Leaf of Langda with a short petiole, slightly wavy margin & acute apex.

Fig. 2 : (a-q) Inflorescence of different varieties of *Mangifera indica*
a. Alphonso – position is terminal, shape pyramidal and colour is green. b. Badshahpasand- position is terminal, shape conical and colour is yellow. c. Goto- position is terminal, shape pyramidal and colour is reddish yellow. d. Sindoria- position is terminal, shape pyramidal and colour is yellow. e. Rucchado- position is extra terminal, shape cylindrical and colour is light green. f. Desi- position is terminal, shape conical and colour is green. g. Dudhpendo- position is terminal, shape pyramidal and colour is red. h. Fasli- position is terminal, shape pyramidal and colour is red. i. Jamadar- position is terminal, shape conical and colour is yellow. j. Cowasji- position is terminal, shape pyramidal and colour is yellow. k. Kesar- clustered inflorescence from same point, position is terminal, shape pyramidal and colour is reddish yellow. l. Khodi- position is terminal, shape pyramidal and colour is yellowish green. m. Totapuri- position is terminal, shape pyramidal and colour is yellow. n. Pairi- position is terminal, shape pyramidal and colour is reddish yellow. o. Rajapuri- clustered inflorescence from same point, position is terminal, shape conical and colour is green. p. Sopari— position is axillary, shape conical and colour is reddish yellow. q. Langda- position is extra terminal, shape pyramidal and colour is yellow.

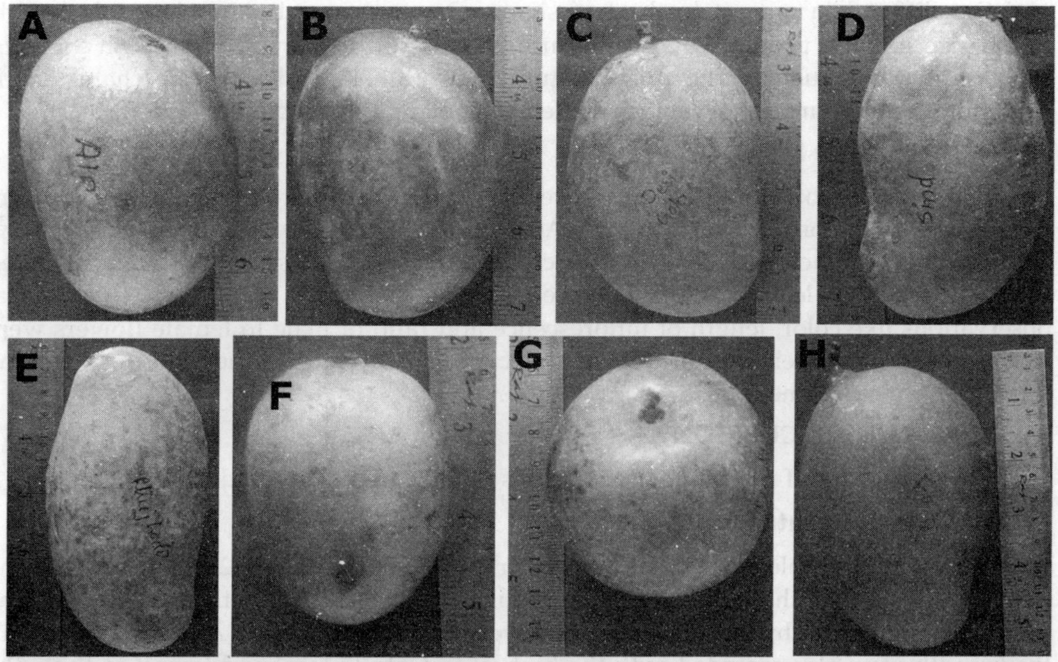

Fig. 3 : (A-Q) Fruit (stage III) of different varieties of *Mangifera indica*.
A. Alphonso- colour is green with red blush on shoulder B. Badshahpasand- colour is green C. Goto- colour is green. D. Sindoria- colour is green. E. Rucchado- colour is green, surface warty. F. Desi- colour is green with yellow blush shoulder. G. Dudhpendo- colour is green H. Fasli- colour is green. I. Jamadar - colour is green. J. Cowasji - colour is green with red blush on shoulder & rough surface. K. Kesar – colour is green. L. Khodi – colour is green. M. Totapuri - colour is green. N. Pairi- colour is green with red blush on shoulder. O. Rajapuri - colour is green. P. Sopari - colour is green. Q. Langda - colour is green.

varieties differed in respect to the leaf shape, size, pulvinus base and petiole length, inflorescence shape, position & colour, bearing habit, fruit size, fruit colour, flesh contents, pulp colour and quality. The present study was aimed to identify the characteristic morphological features of seventeen varieties of mango available in Gujarat.

MATERIALS AND METHODS

The samples of leaves, inflorescence and fruits for the present investigation were collected from the orchard of Junagadh Agricultural University, Junagadh during May-2008.Five trees of each varieties were selected for the study. Observations were recorded by collecting fifty leaves, five inflorescences and five fruits from each tree. Visual observations like color, length of inflorescence and ratio of male to female flowers were made on the field itself. Samples were collected and brought to the laboratory for the quantitative measurement.

RESULTS AND DISCUSSION

The summary of morphological features of the varieties studied are given in table1.

Morphological features

Leaves are simple with no stipules, alternately arranged 13 to 25 cm in length. The petiole varies in length and always with a pulvinate base. The length of the leaves was found highest in Dudhpendo (40cm) and lowest in Neelum (13cm). A great variation in the breadth were also recorded with the highest in Goto (7.1cm) and lowest in Fasli (2.6 cm) The phyllotaxy is usually 3/8 but arranged very closely at the tips which appear to be whorled one.The internode length showed variation, some of the varieties have very short internode giving a clustered appearance. The lamina of some leaves showed wavy margin which were distinct from the non wavy ones. Varieties like Alphonso, Badshahpasand, Goto, Sindoria, Desi, Fasli, Jamadar, Cowasji, Kesar, Khodi, Totapuri, Pairi, Rajapuri, Sopari and Langda showed wavy margins, while Rucchado & Dudhpendo showed smooth margins (fig1). Petiole Length also varied with Dudhpendo showing the highest (4.24 cm) and Langda with the lowest (1.20 cm).

Variations were conspicuously noticed in the inflorescence. The color ranges from green to red. Green colored inflorescence was observed in Alphonso, Rucchado, Desi, Khodi & Rajapuri. Red to yellow colour was observed in Badshahpasand, Goto, Sindoria, Dudhpendo, Jamadar, Kesar, Totapuri, Pairi, Sopari & Langdo. The inflorescences were all triangular or pyramidal/conical in form except for Rucchado which was cylindrical.. The shape of inflorescence was pyramidal in Alphonso, Goto, Sindoria, Dudhpendo, Fasli, Cowasji, Kesar, Khodi, Totapuri, Pairi & Langda while the shape was conical in varieties like Badshahpasand, Desi, Jamadar, Rajapuri & Sopari The length of the internodes in the inflorescence varied. The length of the inflorescence was the highest in Pairi (38-40 cm) and the lowest in Alphonso (10-12 cm). As per the classification based on the density of flower in the inflorescence of Alphonso, Badshahpasand, Rucchado, Dudhpendo, Cowasji, Kesar, Totapuri & Pairi can be categorized into densely flowered. Varieties Goto & Sindoria were in laxly flowered group. Varieties like Desi, Fasli, Jamadar, Khodi, Rajapuri, and Sopari & Langda had medium density of flower in the inflorescence. (fig. 2). All the inflorescence showed two types of flowers, hermaphrodite and staminate flowers. The ratio between bisexual and male flower showed variation. In Pairi, Ruchhado, Sindoria and Sopari the number of bisexual flowers were very high compared to the male flowers. The ratio of bisexual to male was around 90:10. In Alphonso and Khodi the ratio was 40:60 respectively. The highest value of percentage bisexual flowers was recorded in Pairi (95 %) and the lowest in Alphonso (40 %). The highest percentage of male flowers was recorded in Alphonso (60 %) and the lowest in Goto and Sindoria (5 %). Alphonso had 428 male flowers per panicle (Ulhaiah et.al., 1998) and 156 per inflorescence (Anila & Radha, 2003).

The fruit taken for the study was of stage (III) which is the stage just before fruiting.Charaters like length; diameter, weight and colour were studied and are represented in table 1. The maximum weight was of Rajapuri(440g), Cowasji (436.66g), Fazli (355g) Totapuri (352.50g), Kesar (305g), Goto (282.50g), Pairi (245g), Badshahpasand(228.33g), Langdo(226.60g), Jamadar(200g), Rucchado(176.66g) , Alphonso(153.33g), Dudhpendo (135g), Jhumakhiya (125g), Desi (123.33g). Khodi (72.50g) and Sopari (48.33g) showed the lowest weight among the selected varieties. A colour variation in fruits was also seen among the varieties Alphonso, Cowasji & Pairi showed red blush on the shoulder while the other varieties were purely green. (Fig 3). The surface of all the seven varieties is smooth except in Ruchhado (fig 3, E) which showed warty surface. This was due to the presence of very dense fiber strands. The name Ruchhado in vernacular language in Gujarat indicates the presence of large amount of fibers in the fruit pulp.

The present investigation has established that the morphological characters can be used as a tool for the taxonomic discrimination of seventeen varieties of *Mangifera indica.* An artificial synoptic key to the varieties formulated, on the basis of the morphological characters is presented below: (Key 1-3)

Key 1

Identification based on leaf characters

1. Leaf margin slightly undulate- Alphonso, Fasli, Jamadar, Cowasji, Totapuri, Rajapuri, Langda

1.1 Pulvinus base short (0.78-0.98cm)- Fasli, Jamadar

1.1.1 Midvein curved———————————————————**Fasli**

1.1.2 Midvein straight————————————————— **Jamadar**

1.2 Pulvinus base long (1-2cm)- Alphonso, Cowasji, Totapuri, Rajapuri, Langda

1.2.1 Lamina long- (20-25cm)- Totapuri, Alphonso, Cowasji,

1.2.1.1 Leaf apex acute- Cowasji, Totapuri

1.2.1.1.1 Secondary vein at an angle of 45°

1.2.1.1.1.1 Lamina length (24-26cm) ——— **Cowasji**

1.2.1.1.1.2 Lamina length (20-23cm) ——— **Totapuri**

1.2.1.2 Leaf apex acuminate

1.2.1.2.1 Secondary vein horizontal almost at 90°.————————**Alphonso**

1.2.2.1 Lamina short (16-18cm) - Rajapuri, Langda

1.2.2.1.1 Midvein straight

Lamina base equal——————————**Rajapuri**

1.2.2.1.2 Midvein curved

Lamina base unequal——————————**Langdo.**

2. Leaf margin highly undulated- Badshahpasand, Goto, Sindoria, Desi, Kesar, Khodi, Pairi, Sopari

2.1 Interlaminar region embossed——————————**Goto.**

(Interlaminar- laminar region between two secondary lateral veins)

2.2 Interlaminar region unembossed- Badshshpasand, Sindoria, Desi, Kesar, Khodi, Sopari

2.2.1 Lamina base unequal——————————**Desi**

2.2.2 Lamina base equal- Badshahpasand, Sindoria, Kesar, Khodi, Pairi, Sopari

2.2.2.1 Hypopodium long (5cm) —————————— **Badshahpasnd**

2.2.2.2 Hypopodium medium (3-4cm)- Sindoria, Kesar, Khodi, Pairi

2.2.2.2.1 Midvein straight——————————**Khodi**

2.2.2.2.2 Midvein curved ('S' shaped) - Sindoria, Pairi, Kesar

2.2.2.2.2.1 Lamina length 30cm

Breadth 4-5cm——————————**Kesar**

2.2.2.2.2.2 Lamina length 20cm

Breadth 5-6cm——————————**Sindoria**

2.2.2.2.2.3 Lamina length 24cm

Breadth 6-7cm——————————**Pairi**

2.2.2.3 Hypopodium short (2-3cm) —————————— **Sopari**

3. Leaf margin entire- Rucchado, Dudhpendo

3.1 Length of petiole (4-5cm)

3.1.1 Length of Lamina (39-40cm) ————**Dudhpendo**

3.2 Length of the petiole (2-3cm)

3.2.1 Length of Lamina (17-18cm) ————**Rucchado**

Key 2

Identification based on inflorescence characters

1. Inflorescence arising in clusters- Kesar, Rajapuri

1.1 Peduncle reddish yellow

1.1.1 Length of inflorescence (33-35cm) ————**Kesar**

1.2 Peduncle green

1.2.1 Length of inflorescence (22-25cm) ——**Rajapuri**

2. Inflorescence arising solitary-Alphonso, Badshahpasand, Goto, Sindoria, Rucchado, Desi, Dudhpendo, Fasli, Jamadar, Cowasji, Khodi, Totapuri, Pairi, Sopari, Langdo

2.1 Length of inflorescence (12- 15cm) - Alphonso, Cowasji,

2.1.1 Ratio of B: M flowers 40: 60 ——————————————**Alphonso**

2.1.2 Ratio of B: M flowers 20:80——————————————**Cowasji**

2.2 Length of inflorescence (20-25cm)-Badshahpasand, Goto, Fasli, Totapuri, Langda

2.2.1 Ratio of B:M flowers 60:40——————————**Badshahpasand**

2.2.2 Ratio of B:M flowers 90:10——————————————**Langda**

2.2.3 Ratio of B:M flowers 40:60-Fasli, Totapuri

2.2.3.1 Colour of peduncle red——————————————**Fasli**

2.2.3.2 Colour of peduncle yellow——————————————**Totapuri**

2.2.4 Ratio of B: M flowers 5:95- Goto, Sindoria

2.2.4.1 Inflorescence with lateral branches horizontal and long with large internodes **Goto**

2.2.4.2 Inflorescence with lateral branches at an angle of 45°-Sindoria, Sopari

2.2.4.2.1 Internodal length very short——————————————**Sindoria**

2.2.5 Ratio cf B: M flowers 85:15——————————————**Sopari**

2.3 Length of inflorescence (28-32cm) - Rucchado, Desi, Dudhpendo, Khodi

2.3.1 Shape of inflorescence decurrent——————————————**Rucchado**

2.3.2 Shape of inflorescence excurrent- Desi, Dudhpendo, Khodi

2.3.2.1 Ratio of B: M flowers 20:80——————————————**Desi**

2.3.2.2 Ratio of B: M flowers 50:50- Dudhpendo, Khodi

2.3.2.2.1 Colour of peduncle yellowish green——————————**Khodi**

2.3.2.2.2 Colour of peduncle red——————————————**Dudhpendo**

2.4 Length of inflorescence (38- 40cm)- Jamadar, Pairi

2.4.1 Ratio of B: M flowers 70:30

Peduncle reddish yellow——————————————**Pairi**

2.4.2 Ratio of B: M flowers 40:60

Peduncle yellow——————————————**Jamadar**

Key 3

Identification based on Fruit characters

1. Colour of the fruit- Green with red blush on the shoulders- Alphonso, Cowasji, and Pairi

1.1 Fruit shape- Posterior end round- Alphonso, Cowasji

1.1.1 Fruit oval but not elongated

Diameter of the fruit (19- 20cm) ——————————————————**Alphonso**

1.1.2 Fruit oval and elongated (24-25cm) ——————————**Cowasji**

1.2 Fruit shape-Posterior end beaked

Diameter of the fruit (22-23cm) ——————————————————**Pairi**

2. Colour of the fruit- Green-Badshahpasand, Goto, Sindoria, Rucchado, Desi, Dudhpendo, Fasli, Jamadar, Kesar, Khodi, Totapuri, Rajapuri, Soapri, Langda

2.1 Length of the fruit- 5-8cm –Sopari, Khodi

2.1.1 Fruit elongated & oval shaped with narrow posterior beak—**-Khodi**

2.1.2 Fruit rounded with round posterior half——————————**Sopari**

2.2Length of the fruit- 10-15 cm – Rucchado, Desi, Dudhpendo, Badshahpasand, Jamadar, Langdo

2.2.1 Fruit spindle shaped with rounded ends

Surface extremely warty——————————————————**Rucchado**

2.2.2 Fruit globular/ round like a ball———————— ——————**Dudhpendo**

2.2.3 Fruit elongated/ oval shaped-Desi, Badshahpasand, Jamadar, Langdo

2.2.3.1 Colour green with yellow blush on the shoulder———— **Desi**

2.2.3.2 Colour lush green – Sindoria, Badshahpasand, Langda

2.2.3.2.1 Elongated fruit with a depression at posterior end giving the fruit 'S' shape——————————————————————**Sindoria**

2.2.3.2.2 Fruit with no depression on the posterior end giving the fruit an oval shape Posterior end broad, anterior end broader and round————**Badshahpasand**

2.2.3.2.3 Fruit elongated and obovate

Anterior end broad narrowing towards posterior ends————**-—Jamadar**

2.2.3.2.4 Fruit elongated, oblong with broad mid region——————**Langda**

2.2.4 Fruit beaked on both sides

Beaks pointed on both sides giving fruit spindle shaped appearance-**Totapuri**

2.2.5 Fruit beaked on one side- Kesar, Goto, Fasli, Rajapuri

2.2.5.1 Posterior end beaked- Kesar, Rajapuri

2.2.5.1.1 Fruit surface smooth with white conspicuous lenticel———**Kesar**

2.2.5.1.2 Fruit surface glossy with inconspicuous lenticel——————**Rajapuri**

2.2.5.2 Anterior end beaked- Fasli, Goto

2.2.2.2.1 Tapering beak at the anterior end, posterior end round—**Fasli**

2.2.2.2.2 Posterior beak blunt with inconspicuous lenticels—— **Goto**

REFERENCES

Anila, R. and T Radha, 2003. Studies on fruit drop in Mango varieties, *J of Tropical Agriculture* (4) 30-32.

Hossain, M.M., Haque, Rahim, M.A., M.A. and Rahman, M.H. 2001. Physio-Morphological and compositional variation in ripe fruit of three mango varieties. Journal of Biological Sciences. 1(11): 1101-1102.

Majumdar, P.K. Sharma, D.K. and Sanyal, D. Mango. Fruits. Tropical and sun-tropical Fruits. Vol-I. Naya Udyog. Calcutta.

Medlicott, A.P. Bhogal,M and Reynolds,S.B (1986) Changes in Peel pigmentation during ripening if mango fruit. (*Mangifera indica* var. Tommy Atkins). Ann. Appl Biol 109: 651-656.

Muhammad, A.Muhammad, U. Muhammad, J.J and Muhammad, M.K. (2002). Comparative study of flower sex ratio in different cultivars of Mango (*Mangifera indica* L.) International Journal of Agriculture and Biology. 2 : 220-222.

Paul, V. Malik, S.k. Srivastava, G.C. 2007. Intervarietal differences in the surface morphology and anatomy of Mango (*Mangifera indica*) Fruit. Phytomorphology. 57 (3&4): 211-220.

Singh, R.N 1978, Mango. *Indian Council of Agriculture Research*, New Delhi. 39-55.

Sturrock, T.T. (1996) The Mango Inflorescence. Florida State Horticultural Society.366-369.

Zaied, N.S. Khafagy, S.A.A and Saleh, M.A (2007). Evaluation of some Mango species by fruit characters and fingerprint. Research Journal of Agriculture and Biological Sciences. 3 (4): 316-320.

❑❑❑

Chapter 9

Biology, Propagation, Utilization and Economics of *Prosopis cineraria* (Khejri) in the Thar Desert, India

PAWAN K. KASERA*, JITENDRA K. SHUKLA AND SHER MOHAMMED**

INTRODUCTION

Trees of the genus *Prosopis* are important in the development of rural economy of many arid and semi-arid regions of the world. Its remarkable value in the arid zone lies in their adaptation to extreme heat, drought, salinity & alkalinity, drifting sand, grazing and repeated lopping. Species of *Prosopis* are natives to South, Central and North America, Africa and Asia. It has about 44 species, out of these, 40 are native to America and one (*P. africana*) to Africa. The other three old world species, e.g. *P. cineraria*, *P. farcata* and *P. koelziana* are native to the Middle East and Pakistan, with the range extending into India for *P. cineraria* and into Cyprus and subtropical North Africa for *P. farcata* (Fagg and Stewart, 1994).

Historically, *P. cineraria* [(Linn.) Druce; Family: Mimosaceae] known as *Khejri* in N.W. Indian desert has played a significant role in the rural economy of this region of Indian sub-continent. Naturally occurring *Khejri* has been reported from Afghanistan, Iran (Persia), Pakistan (Baluchistan & Sind) and India (Rajasthan, Delhi, Haryana, Gujarat, North Karnataka & Madhya Pradesh) (Puri *et al.*, 1964). In the arid western Rajasthan, the distribution pattern along a transect from Aravallis ranges in the east to Jaisalmer in the west is discontinuous and varied. Its density increases from the western to the eastern part of the western Rajasthan. The rainfall areas between 250 to 400 mm show its high population, but it can also grow in extreme arid tract (100 mm rainfall). Its discontinuous distribution pattern is due to its habitat preferences and specific water

Laboratory of Plant Ecology, Department of Botany, Jai Narain Vyas University, Jodhpur 342 033, India

*Corresponding author, Govt. Lohia (PG) College, Churu 331 001 (Raj.), India

**Lecturer in Botany, Govt. Lohia (PG) College, Churu 331 001 (Raj.), India

requirements. On less favourable habitats the density, height and crown cover decrease than those of preferred habitats (Shankar, 1980). In the present article main emphasis is given on the biology, economic importance, seed germination, depth and time of seed sowing, propagation techniques, manure and fertilizer requirements, water management, intercropping and economics of *P. cineraria*.

BIOLOGY AND ECONOMIC UTILITY

Prosopis cineraria is an important leguminous multipurpose slow growing tree species of the N.W. Indian Thar desert. In nature, it reproduces by seeds only. However, this tree can be raised through air-layering and sprouting of vegetative buds on the mature stem and roots. It has deep root system with low water and nitrogen requirements. Shoots grow to produce new leaves which appear twice in a year, i.e. in March-April and July-October and thereafter these leaves develop slowly. The leaf-fall is observed from November onwards followed by flowering in January, which reaches in full bloom during February-March. The fruit setting starts from April onwards which attains maturity by May.

P. cineraria is a major entity of traditional agroforestry system, which is grown together with field crops to provide dry season fodder and fuel. Its roots penetrate deep into the soil and do not compete with nearby plants for water and minerals, which make it ideal for its introduction in agroforestry system. The desirable features that favour *P. cineraria* in agroforestry are its slow growth, straight stem and less shade effects on the intercrops, deep root system, mono-layered canopy and ability to fix atmospheric nitrogen. Being a multipurpose tree its parts from apex to root including bark are used in one form or the other. The wood is suitable for interior construction work such as column for huts, roofs, doors, windows, wheel and hub of carts, small agriculture implements, tool handles, small turnery articles and well curbs (Anonymous, 1969). Roots contain significant amount of tanning material. The leaves locally called *'loong'* are highly palatable and used as a nutritious top feed for cattle and livestock. The protein rich pods not only provide considerable amount of moisture to the animals during peak summer months of May-June, but to soldiers also in a sandy field who chew leaves to satisfy thirst. A moderately grown tree yields nearly 25-30 kg of dry leaf forage per year (Bohra and Ghosh, 1980). The pods normally used as fodder for livestock before ripening, which are rich in sweetish farinaceous pulp, consumed as food. The pods either green or dry after boiling serve as green and dry vegetable known as *'Sangari'* by rural masses. The dry pods are fed to milch cattle and one full grown tree gives at least 5 kg of ripened pods. The dry pods reduce the quest for water in summer months and generally nomads eat them in dry periods. The powdered inflorescence mixed with sugar and water is used for a cooling effect, as a blood purifier and for prevention of boils and skin diseases (Purohit and Khan, 1980; Singh and Pandey, 1998). The bark is dry, bitter with sharp taste, cooling and anthelmintic, tonic, cures leprosy, dysentery, bronchitis, asthma, leucoderma, piles, tumors of the muscles, wandering of mind and scorpion and snake bites (Kirtikar and Basu, 1994). The gum oozing out in February-March from tree stem is used by rural folks with sweet balls after child delivery, which is considered nutritive. The firewood makes a better charcoal among local plants of the arid zone, which gives better heat. The wood ash contains 31% soluble potassium salts, which may be used as source of potash (Purohit and Khan, 1980). Ashes are rubbed over the body to remove hairs (Memon *et al.*, 1988). Besides increasing soil fertility, it acts as a soil binder and lowers down the speed of hot winds.

The water soluble fraction of the methanol extract of the bark shows significant anti-inflammatory activity (Satyavati *et al.*, 1987). Ukani *et al.* (2000) documented that bark is efficacious in rheumatism. A new alkaloid-spicigerine isolated and characterized as w-(3-hydroxy-2-methyl-6-piperidyl) alkonic acid, fatty acid composition determined as palmitic, stearic, oleic and linoleic acids by GLC; b-sitosterol, galactose, mannose, glucose, arabinose and rhamnose are identified in seeds (Rastogi and Mehrotra, 1991). The petroleum ether extract of the seeds is reported to yield two new flavones, Prosogerin C and Prosogerin D (Bhardwaj *et al.*, 1978, 1980). Khan and Harsh (2000) reported an effective and commercial plant growth regulator, i.e. Triacontanol (0.012%) in the leaves of *P. cineraria*, which is responsible for better grass growth under canopy and in the vicinity of this tree. The stem bark contains vitamin K and n-octacosyl acetate. A cytotoxic principle, i.e. patulibin has been isolated from flowers (Khare, 2007).

SEED GERMINATION BEHAVIOUR

Hard seed coat dormancy inhibits germination in *P. cineraria*. Sulphuric acid scarification is useful to overcome the resistance of hard seed coat to imbibition of water. Seed germination studies were performed in sterilized petri dishes lined with single layer of moistened filter paper. The experiment was carried out in continuous white light (1000 lux) at 28 ± 2° C. The experimental setup was triplicated and each petri plate contained 10 seeds. Seeds soaked in water for 24 h after acid scarification show higher germination than those kept in running tap water for 3-4 h after acid scarification. The maximum (100.0%) and minimum (26.67%) germinations were obtained in seeds scarified for 15 min and non-scarified seeds soaked in water for 24 h, respectively. The maximum root and shoot lengths were observed in 15 and 5 min acid scarified seeds, respectively. The data were significant at 5% and 1% probability levels (Table 1). Shukla (2001) had earlier observed that 15-20 min conc. sulphuric acid scarified seeds exhibited higher germination percentage.

The differences in shape, size, volume and weight of seeds have been found to play an important role in regulation of seed germination. Manga (1995) and Shukla *et al.* (2000a) reported variability for various seed traits, viz. weight, length, width, thickness, volume and density. They have concluded that seed weight, thickness and volume associated positively and significantly with percent germination and other germination parameters. Seed thickness and volume have higher direct effect on germination percentage as compared to weight.

DEPTH AND DURATION OF SEED SOWING

Sowing depths of seeds play an important role in germination. Along with seed weight, sowing at optimum depth is an important aspect to obtain higher germination and good quality seedlings (Chandra and Atma Ram, 1980). As the moisture is generally a limiting factor in arid environment, the placement of seeds at proper depth in nursery is essential to provide the most favourable environment for uniform germination and seedling growth. Seeds were sown at four different depths (1, 2, 3 and 4 cm) in ten replicates and each polybag contained two seeds. Maximum seedling emergence takes place when seeds are sown at 2 cm depth (Table 2).

TABLE 1

Effect of different durations of conc. H_2SO_4 acid scarification on seed germination (%) and seedling growth (cm) of *P. cineraria* (observations taken after 10 days)

Durations (min)	Germination		Seedling growth			
			Root length		Shoot length	
	A	B	A	B	A	B
0	30.00	26.67	1.41	1.63	0.93	1.18
5	70.00	66.67	1.66	1.33	2.64	1.34
10	86.67	66.67	1.78	1.32	2.52	1.39
15	96.67	100.00	2.47	1.92	2.24	2.02
20	83.33	93.33	1.90	2.45	2.58	2.18
CD	20.48**	43.21*	1.00*	1.24*	0.64**	0.84*

= Seeds kept in running tap water for 3-4 h after acid scarification.

B = Seeds soaked in tap water for 24 h after acid scarification.

* & ** p = Significant at 0.05 & 0.01, respectively.

(Data were analyzed statistically as per Gomez and Gomez, 1984).

TABLE 2

Effect of different depths and durations of seed sowing on seedling emergence in *P. cineraria* under nursery conditions

Seed sowing		Seedling emergence	
		Days	Percent
Depth (cm)	1.0	3	70.0
	2.0	4	90.0
	3.0	6	60.0
	4.0	8	35.0
	CD	1.73**	4.99**
Duration (date)	1st May	4	65.0
	11th May	3	65.0
	21st May	3	70.0
	1st June	3	75.0
	11th June	3	70.0
	21st June	3	60.0
	CD	NS	NS

NS = Non-significant; and ** p = Significant at 0.01.

Periodicity and peak period of germination are important characteristics of seeds, which make seedlings appear in field at regular successions and season after season. Climatic conditions in any one year may favour germination of one species more than the others. Freshly collected seeds often show very poor germination in natural field conditions and better germination in controlled conditions. Investigations have shown that *P. cineraria* seeds showed maximum germination during last ten days of May to first ten days of June under nursery conditions (Table 2). In general, Sen (1982) reported that freshly collected and stored seeds need to be sown when the environmental conditions are congenial for them. Seeds of *Salvadora persica*, an arid zone species exhibited maximum germination when they were sown during 1st & 2nd week of June under nursery conditions (Prakash, 2001).

PROPAGATION TECHNIQUES

Plant propagation in nature is an important aspect for its sustenance. *P. cineraria* can be easily propagated by seeds after overcoming hard seed coat dormancy, besides air-layering. However, propagation by cuttings yielded only shoot sprout but no root formation. The plants can be multiplied through seeds primarily or through air-layering.

Seeds: The acid scarified seeds when sown in polybags containing mixtures of sand: clay: FYM in 1:1:1, 1:2:1, 2:2:1 and 2:1:2 ratios under nursery conditions. 30 replicates were maintained in each soil mixture ratio. Maximum plant height, collar diameter, root length and biomass were observed in 2:2:1 ratio (Table 3).

TABLE 3

Effect of different soil mixtures on growth parameters (cm) and plant biomass (g plant^{-1} dry wt.) of *P. cineraria* after six months under nursery conditions

Soil mixtures	Growth parameters			Biomass	
(Sand:clay:FYM)	Plant height	Collar diameter	Root length	Aboveground	Belowground
1:1:1	50.7	0.31	47.0	1.55	1.28
1:2:1	47.5	0.32	47.3	1.44	0.91
2:2:1	53.0	0.36	60.0	1.99	1.39
2:1:2	45.5	0.33	60.0	1.47	1.36
CD	1.53**	NS	1.0**	0.01**	0.12**

NS = Non-significant; and ** p = Significant at 0.01.

Air-layering: Vegetative propagation is opted because of its various advantages such as maintenance of particular characteristics of the plant adaptability to a particular habitat, development of resistance to pests and diseases and drought tolerance to modify the growth of plant. Air-layering is the uncommon method of vegetative propagation in which root formation is an important aspect.

Air-layering experiments were conducted during rainy season on 8-10 years old trees by removing the stem bark (3-4 cm) in a girdle-like manner from the internodal portion with the help of a knife and wrapped with moist clay and covered with layers of gunny bag followed by polythene sheet and tied properly with thread. The soil on the girdled portion was kept moist with water as far as possible. This arrangement was continued for 45-50 days for initiation and development of roots. The rooted air-layered stem branches were cut and removed from mother plants and transplanted in pits containing soil mixture of sand: clay: FYM (2:2:1) after removal of polythene.

IBA treated branches exhibited maximum root formation and success in field as compared to other treatments (Table 4). The plants raised through air-layering technique showed higher values of growth parameters, biomass, HI and early flowering and fruiting as compared to seed raised plants (Table 5).

TABLE 4

Effect of different treatments on root parameters and success in air-layered branches of *P. cineraria* under field conditions

Treatments	Rooting (%)	Number of roots	Max. root length (cm)	Success (%)
Control	50.0	18	5.0	35.7
IBA (10 mg l^{-1})	80.0	30	4.5	50.0
Puncture by needle	75.0	25	4.0	0.0
CD	6.54**	1.31**	NS	13.09**

NS = Non-significant; and ** p = Significant at 0.01.

TABLE 5

Growth parameters (cm), plant biomass (kg ha^{-1} dry wt.) and Harvest Index (HI; %) in one-year-old plants of *P. cineraria* raised through air-layering technique and seeds

Propagation techniques	Growth parameters			Biomass		HI
	Plant height	Collar diameter	Root length	Above-ground	Below-ground	
Air-layering	94.0	1.76	100.0	62.40	21.60	6.19
Seeds	74.6	1.04	68.0	4.92	2.69	2.61
CD	12.42*	0.248**	12.42**	116.76*	19.88**	0.012**

* and ** p = Significant at 0.05 and 0.01, respectively.

Prakash *et al.* (2001) reported that air-layering was the fastest method and is 100% successful for inducing rooting in *Commiphora wightii*, an arid zone plant. IAA and Seradix B-3 proved effective for initiation of roots over control in *P. cineraria* (Rawat *et al.*, 1982; Solanki *et al.*, 1986). Felker and Clark (1981) reported 30% success in air-layered twigs of *P. cineraria*. Thus for large-scale multiplication, air-layered plants should be treated with IBA (10 mgl^{-1}) during rainy season before transferring in the field to obtain better growth performance, high biomass yield and HI.

SPACING

Maintaining an optimum plant population assumes a great importance in yield maximization. Spacing requirement may vary according to cultivar, soil fertility, soil conditions and also agroclimate of a particular region. With increasing demand for fodder, fuel and timber, tree planting activities are accorded the highest priority in the desert afforestation programmes. *P. cineraria* with narrow spacing that suffix agronomical operations are expected to improve soil biological quality and thus soil fertility (Purohit *et al.*, 2002). Maximum plant height and collar diameter were observed in 5 x 5 m spacing followed by 4 x 4 and minimum in 3 x 3 m (Table 6). The initial spacing of 2 m is generally adopted for *P. cineraria* which may be increased to 6 m later (Anonymous, 1969). Gupta *et al.* (1998) suggested 4 x 6 m spacing was favourable after four years of growth in *P. cineraria*.

TABLE 6

Effect of different spacing regimes on the growth parameters (cm) of *P. cineraria* during different months

Spacing (m)	Growth parameters					
	Plant height (months)			Collar diameter (months)		
	12	24	36	12	24	36
3 x 3	34.0	49.5	56.5	0.41	0.52	0.66
4 x 4	43.0	55.0	60.4	0.55	0.79	0.83
5 x 5	48.4	59.7	73.2	0.52	0.80	1.00
CD	NS	NS	NS	NS	NS	NS

NS = Non-significant.

MANURE AND FERTILIZERS

Various treatments for the maximum yield in the field conditions, viz. arbuscular mycorrhizae (AM) infested soil (100 g plant^{-1}); FYM (farmyard manure; 8-10 t ha^{-1}; 500 g plant^{-1}); hexameal (organic manure; 40 q ha^{-1}; 100 g plant^{-1}); NPK: full dose (30:40:15 kg ha^{-1}; 3.2:11.3:1.25 g plant^{-1}); NPK: half dose (30:20:15 kg ha^{-1}; 3.2:5.65:1.25 g plant^{-1}); NPK: full dose + hexameal and NPK: half dose + hexameal, besides control were provided every month upto one year after one month of establishment of plants under field conditions. For AM treatment, AM infested 100 g soil was mixed with the soil mixture. For this soil was collected underneath two desert grasses, *Cenchrus ciliaris* and *C. setigerus*, containing spores of genus *Glomus*. Each treatment has 30 replicates. BT means the growth parameters of young seedlings before providing any treatment.

Plants in desert are limited by slow growth as compared to those in moist areas. Seedlings planted in prolonged dry habitats are exposed to insufficient moisture and deficient nutrients so that their initial increment in biomass is retarded; consequently leading to reduction in survival percentage. Initial reduction in growth will be reflected on the later stages of growth increment and biomass. Therefore, it is necessary to make nutrients available in the form of organic, inorganic or biofertilizers and adequate supply of water to boost the growth of seedlings at the initial stages for better survival. The application of NPK: full dose + hexameal in *P. cineraria* favourably influences the plant height and collar diameter followed by those in association of arbuscular mycorrhizae after 24 months of growth period (Table 7). The maximum biomass is achieved in NPK: full dose + hexameal treated plants followed by FYM. The bark yield and HI & leaf yield increased significantly in NPK: full dose + hexameal and FYM treated plants, respectively (Table 8). Thus, the treatments of NPK: full dose in combination with hexameal is found beneficial for optimum growth and biomass production, besides AM and FYM.

TABLE 7

Effect of different treatments on growth parameters (cm) of *P. cineraria* after 24 months under field conditions

Treatments	Growth parameters				% increase after 24 months	
	Plant height		Collar diameter			
	BT	24	BT	24	Plant height	Collar diameter
Control	27.5	62.0	0.198	0.908	125.45	358.59
AM	19.8	82.2	0.210	1.252	315.15	500.00
FYM	21.5	86.8	0.180	1.022	303.72	467.78
Hexameal	22.1	58.0	0.181	0.763	162.44	321.55
NPK: full dose	21.3	61.6	0.198	0.627	189.20	216.66
NPK: half dose	18.9	52.7	0.176	0.843	178.84	378.98
NPK: full + hex.	22.5	103.7	0.190	1.260	360.89	558.94
NPK: half + hex.	25.1	89.2	0.200	0.853	255.38	326.50
CD	NS	19.931**	NS	0.379*		

BT = Before treatment; hex. = Hexameal; NS = Non-significant; and * & ** p = Significant at 0.05 & 0.01, respectively.

With combined applications of N and P to *P. cineraria*, Gupta *et al.* (1996) obtained improved crown and collar diameter and plant height. Gupta *et al.* (2000) also observed maximum increment of plant height in *Ziziphus mauritiana* by an application of FYM. 15% increase in seedling height of *P. cineraria* by application of FYM was reported by Gupta (1992). Tiwari and Pandey (2000) observed maximum growth by application of 12 kg FYM plant^{-1} under irrigated conditions. Shukla *et al.* (2000b) found that collar diameter and HI increased significantly in *P. cineraria* after applications of hexameal and FYM, respectively.

WATER MANAGEMENT

The plants under different irrigation schedules were initially watered as per the requirement till their establishment. Growth and development of plants in arid regions are dependent largely on the monsoon rain, which is highly unpredictable. Irrigation is a costly input in crop production and has not received adequate attention of Indian researchers working on medicinal and aromatic plants (Gupta and Chadha, 1995).

TABLE 8

Effect of different treatments on biomass, leaf & bark yields (q ha^{-1}dry wt.) and Harvest Index (HI; %) of *P. cineraria* after one year of growth under field conditions

Treatments	Biomass		Leaf yield	Bark yield	HI
	Above-ground	Below-ground			
Control	0.286	0.286	0.091	0.020	0.089
AM	0.501	0.102	0.457	0.007	0.031
FYM	1.953	1.454	0.570	0.310	0.228
Hexameal	0.155	0.116	0.077	0.010	0.093
NPK: full dose	0.340	0.447	0.108	0.029	0.091
NPK: half dose	0.936	0.726	0.249	0.051	0.076
NPK: full + hexameal	2.421	1.635	0.406	0.358	0.221
NPK: half + hexameal	1.803	0.932	0.545	0.088	0.081
CD (p = 0.01)	0.145	0.125	0.075	0.029	0.018

After rainy season, the irrigation (8 l plant^{-1}) was provided fortnightly and at monthly interval under field conditions. Monthly irrigation favourably influenced growth parameters as well as plant biomass and fortnightly irrigation for HI under field conditions in *P. cineraria* (Table 9).

TABLE 9

Effect of different irrigation schedules on the growth parameters (cm), biomass, bark yield (kg ha^{-1} dry wt.) and Harvest Index (HI; %) of *P. cineraria* after 12 months

Irrigation Schedules	Growth parameters		Biomass		Bark		HI
	Plant height	Collar diameter	Above-ground	Below-ground	Fresh	Dry	
Control	41.83	0.41	4.78	4.87	3.60	0.54	5.59
Fortnightly	56.08	0.66	9.71	6.87	5.60	2.06	12.42
Monthly	71.33	1.03	38.93	21.07	11.64	6.26	10.44
CD (p = 0.01)	13.70	0.10	19.37	6.80	0.38	0.54	1.18

Gupta and Sharma (1998) advocated water harvesting as a technique, which helps in early growth and establishment of tree seedlings. Beneficial effect of different irrigation levels on growth and biomass potentials of *Leucaena leucocephala* was reported by Singh *et al*. (2000). Irrigated *P. cineraria* plants exhibited better growth as compared to unirrigated ones (Tiwari and Pandey, 2000).

TABLE 10

Growth (cm) and yield (q ha^{-1}) of pearl millet grown as sole crop and intercrop with *P. cineraria* (adapted after Kasera *et al*., 2002)

Cropping systems	Pearl millet					*P. cineraria*	
	Plant height	No. of nodes	Ear diameter	Grain yield	Fodder yield	Plant height	Collar diameter
Sole crop	157.40	6	2.52	21.50	224.99	-	-
P. cineraria	-	-	-	-	-	94.60	2.155
Pearl millet + *P. cineraria*	192.80	8	3.06	30.28	424.99	118.50	2.319
CD	28.629*	1.110*	0.207**	3.991*	146.432*	NS	NS

NS = Non-significant; and * & ** p = Significant at 0.05 & 0.01, respectively.

TABLE 11

Growth (cm) and yield (q ha^{-1}) of cluster bean grown as sole crop and intercrop with *P. cineraria* (adapted after Kasera *et al*., 2002)

Cropping systems	Cluster bean				*P. cineraria*	
	Plant height	No. of branches	No. of clusters	Grain yield	Plant height	Collar diameter
Sole crop	84.00	4	3	8.65	-	-
P. cineraria	-	-	-	-	94.60	2.155
Cluster bean + *P. cineraria*	101.4	5	4	12.95	108.40	2.316
CD	16.295*	NS	NS	2.845*	NS	NS

NS = Non-significant; and * p = Significant at 0.05.

INTERCROPPING

The intercropping experiments were studied by pruning 3 years old plantation of 2x2 m spacing. The experiment was laid in randomized block design with three replications. The size of plot was 10x2 m. Cluster bean (guar; var. *Maru*) and pearl millet (bajra; var. *Eknath*) were sown @ 20 kg ha^{-1} maintaining a spacing of 30x15 cm and @ 5 kg ha^{-1} maintaining 45x10 cm spacing, respectively. Intercropping on growth performance of *P. cineraria* in 3 years old pruned plants revealed that cluster bean and pearl millet between *P. cineraria* are beneficial (Tables 10 & 11). A highly synergistic relationship between *P. cineraria* and agricultural crops (pearl millet & cluster bean) was obtained when grown together. Plant height, number of nodes, ear diameter, grain and fodder yield of pearl millet was significantly higher when grown with *P. cineraria* over sole crop (Table 10). Similarly, the data in Table 11 on cluster bean show a significant increase in plant height and grain yield and non-significant increase in the number of branches and clusters when grown with *P. cineraria*.

P. cineraria tree has a boosting effect on the yield of crops growing in its vicinity. The organic matter, nitrogen, potassium and phosphorus were high under its canopy (Singh and Lal, 1969), probably due to more vigorous growth of vegetation thus producing more litter, which is mineralized rapidly as canopy provides favourable moisture conditions (Gupta and Saxena, 1978).

ECONOMICS

The net return was calculated by subtracting cost of cultivation from the economic yield. B: C ratio was computed by dividing gross return with cost of cultivation. For maximum utilization of the medicinal plant diversity of arid region, there is a need to work out their cost-benefit (B: C) ratio cultivated in this region. Data from Table 12 show that cultivation of *P. cineraria* gave the net profit of Rs. 22500/- ha $^{-1}$ with B: C ratio of 2.65:1 after two years of plantation. Other crops can also be cultivated as intercrops and additional returns can be realized.

TABLE 12

Cultivation economics in *P. cineraria*

Items of expenditure	Amount (Rs. ha^{-1})
A. Expenditures	
Cost of raising of nursery and transplanting	1000.00
Land preparation and layout	1500.00
Cost of FYM (farmyard manure) and soil mixtures	1500.00
Weeding and hoeing	1000.00
Irrigation	1500.00
Harvesting (labour charges)	2000.00
Total	**8500.00**
B. Returns	
Bark yield (500 kg ha^{-1}; @ Rs. 50 kg^{-1})	25000.00
Leaf yield (1500 kg ha^{-1}; @ Rs. 4 kg^{-1})	6000.00
Total	**31000.00**

C. Net profit per hectare (31000-8500) = Rs. 22500.00

D. B: C ratio: 2.65:1

INSECTS AND PESTS

Very little information is available about the insect-plant association in *P. cineraria* (Parihar, 1993).

Gall forming insects: Gall formation reduces vegetative growth and seed formation in *P. cineraria.* The genesis of these growths is a result of a nutritional dependence of the causal organisms on plant tissues. Gall on branches, leaflets and inflorescence are reported to be caused by *Pediobopsis* spp., *Contarinia prosopidis, Eriophyes prosopidis* and *Assura albicostalis.*

Defoliators: Chafers or Scarabacid beetles constitute the most prominent group of defoliators. The more common Scarabacids are: *Lachnosterna consanguinea, Anomala* spp., *Adoretus* spp. and *Schizonych ruficollis*. Locusts (*Schistocerca gregaria*) severely damage the foliage when its invasions occur.

Sap suckers: More important sap sucking insects are *Oxyrhachis tarandus, O. rufescens* and *Homoeocenus* spp. Application of methyl parathion dust @ 20 kg ha $^{-1}$ is helpful for controlling all these pests.

GENETIC IMPROVEMENT

P. cineraria is a slow growing tree and for its improvement knowledge of genetic parameters such as heritability, genetic gain, etc. are prerequisites. Mass selection for improvement of tree height is better than family selection, but there is no control of pollen parents in mass selection. Combined selection, i.e. selecting the best trees by using a large number of families will be most effective for improvement of tree height (Jindal and Singh, 2006).

SUMMARY

Studies reveal that *Prosopis cineraria* is a multipurpose tree species of the north western Indian desert. However, the people of this region are ignorant with its medicinal properties, which need to be exploited. The slow growth of this species can be improved to some extent by air-layering technique utilizing 10 mg l^{-1} of Indole butyric acid (IBA) solution. For large-scale cultivation, it can be multiplied through seeds after acid scarification, as they have hard coat dormancy. 15 minutes conc. sulphuric acid scarified seeds when sown after 3-4 h washing in running tap water during the last week of May to first week of June at 2 cm depth under nursery conditions give maximum germination. The soil mixture ratio (2:2:1) of sand: clay: FYM favours the optimum growth under nursery and field conditions. Irrigation at monthly interval is sufficient for optimum growth. Because of its ameliorating properties and less shade effect, it is an ideal component of agri-silvicultural practices. Its cultivation gives a net profit of Rs. 22500/- ha $^{-1}$ with cost: benefit (B: C) ratio of 2.65:1 after two years of plantation.

REFERENCES

Anonymous 1969. *The Wealth of India*. Vol. III, CSIR, New Delhi: 247-249.

Bharadwaj, D.K., Bisht, M.S., Jain, R.S. and Sharma, G.C. 1980. Prosogerin-D, a new flavon from *Prosopis spicigera* seeds. *Phytochemistry* **19**: 1269-1270.

Bharadwaj, D.K., Jain, R.K., Sharma, G.C. and Mehta, C.K. 1978. Prosogerin-C, a new flavon from *Prosopis spicigera* seeds. *Indian J. Chem.* **16B**: 1133.

Bohra, H.C. and Ghosh, P.K. 1980. The nutritive value and digestibility of Loong [*P. cineraria* (Khejri)] leaves. In: *Khejri (Prosopis cineraria) in the Indian Desert - Its Role in Agroforestry* (Mann, H.S. and S.K. Saxena, eds). CAZRI Monograph No. 11, CAZRI, Jodhpur, pp. 45-50.

Chandra, J.P. and Ram, Atma 1980. Studies on depth of sowing deodar seed. *Indian Forester* **106**: 852-855.

Fagg, C.W. and Stewart, J.L. 1994. The value of *Acacia* and *Prosopis* in arid and semi-arid environment. *Jour. Arid Environ.* **27**: 3-25.

Felker, P. and Clark, P.R. 1981. Rooting of mesquite (*Prosopis*) cuttings. *Jour. Range Manag.* **34**: 466-468.

Gomez, K.A. and Gomez, A.A. 1984. *Statistical Procedures for Agricultural Research* (2nd Ed). John Wiley & Sons, New York, USA.

Gupta, G.N. 1992. Influence of different soil mixture on nursery growth of some arid zone tree species. *Indian Forester* **118**: 952-960.

Gupta, G.N., Singh, G. and Kachwaha, G.R. 1998. Performance of *Prosopis cineraria* and associated crops under varying spacing regimes in the arid zone of India. *Agroforestry Systems* **40**: 149-157.

Gupta, G.N., Kuppusamy, V., Choudhary, K.R., Bohra, N.K., Singh, N. and Lata, K. 1996. Effect of fertilizers on early growth of desert species: *Prosopis cineraria* and *Tecomella undulata*. *Annals Forestry* **4**: 29-33.

Gupta, J.P. and Sharma, A.K. 1998. Integrated effects of water harvesting and manuring on growth and establishment of *Prosopis cineraria* (Khejri) in hot arid region of Rajasthan. *Indian Forester* **124**: 54-58.

Gupta, J.P. and Saxena, S.K. 1978. Studies on the monitoring of the dynamics of moisture in the soil and the performance of ground flora under desertic communities of trees. *Indian J. Ecol.* **5**: 30-36.

Gupta, J.P., Ojasvi, P.R. and Goyal, R.K. 2000. Rehabilitation of degraded land in arid ecosystem: a case study. In: *Environmental Protection* (Choudhary V., K. Singh and B.L. Kakralya, eds.). Pointer Publisher, Jaipur, pp. 215-223.

Gupta, R. and Chadha, K.L. 1995. Medicinal and aromatic plants research in India. In: *Advances in Horticulture: Medicinal and Aromatic Plants* (Chadha, K.L. and R. Gupta, eds.). Vol. 11, Malhotra Publishing House, New Delhi, pp. 1-43.

Jindal, S.K. and Singh, M. 2006. Estimation of components of variance, heritability, family selection versus mass selection for tree height in *Prosopis cineraria*. *Annals Arid Zone* **45**: 183-187.

Kasera, P.K., Naruka, I.S. and Shukla, J.K. 2002. Interaction of young Khejri (*Prosopis cineraria*) trees with the growth and yield of pearl millet and cluster bean intercrops. *My Forest* **38**: 167-170.

Khan, H.A. and Harsh, L.N. 2000. Role of minor forest products (MFP) producing plants in Governmental protection. In: *Environmental Protection* (Choudhary V., K. Singh and B.L. Kakralya, eds.). Pointer Publisher, Jaipur, pp. 240-243.

Khare, C.P. 2007. *Indian Medicinal Plants: An Illustrated Dictionary*. Springer International Edition, Rajkamal Electric Press, Delhi.

Kirtikar, K.R. and Basu, B.D. 1994. *Indian Medicinal Plants.* Bishen Singh Mahendra Pal Singh, Dehradun, Vols. 1-4: 910-912.

Manga, V.K. 1995. Diversity and inter-relationship study in seed and seedling traits of *Prosopis cineraria* (L.) MacBride. Ph.D. Thesis, J.N. Vyas University, Jodhpur, India.

Memon, M.I.A., Shahani, N.M. and Syed, G.M. 1988. *Glimpses of Medicinal Plants of Sind*. Sind Agriculture Univ., Tandojam Sind, Pakistan: 79-80.

Parihar, D.R. 1993. Insect fauna of Khejri, *Prosopis cineraria* of arid zone. *Indian Jour. Forestry* **16:** 132-137.

Prakash, J. 2001. Biology of some medicinal plants in Indian arid zone. Ph.D. Thesis, J.N. Vyas University, Jodhpur, India.

Prakash, J., Kasera, P.K. and Chawan, D.D. 2001. Multiplication of *Commiphora wightii* through air layering technique in Indian desert. *Sci. & Cult.* **67**: 65-67.

Puri, G.S., Jain, S.K., Mukharjee, S.K., Sarup, S. and Kotwal, N.K. 1964. Flora of Rajasthan. *Rec. Bot. Surv. India* **19**: 1-159.

Purohit, M.L. and Khan, W. 1980. Socio-economic dimensions of Khejri [*Prosopis cineraria* (Linn.) MacBride]. In: *Khejri (Prosopis cineraria) in the Indian Desert - Its Role in Agroforestry* (Mann, H.S. and S.K. Saxena, eds). CAZRI Monograph No. 11, CAZRI, Jodhpur, pp. 56-63.

Purohit, U., Mehar, S.K. and Sundaramoorthy, S. 2002. Role of *Prosopis cineraria* on the ecology of soil fungi in Indian desert. *Jour. Arid Environ.* **52**: 17-27.

Rastogi, R.P. and B.N. Mehrotra (1991). *Compendium of Indian Medicinal Plants,* (eds.). Vol. II, CDRI, Publication & Information Directorate, New Delhi.

Rawat, T.S., Sharma, H.C. and Pundir, J.P.S. 1982. Air layering a success in propagation of Khejri (*Prosopis cineraria* L.). *Udyanika* **5**: 43-44.

Satyavati, G.V., Gupta, A.K. and Tandon, N. 1987. *Medicinal Plants of India,* (eds.). Vol. 2, ICMR, New Delhi: 504-509.

Sen, D.N. 1982. *Environment and Plant Life in Indian Desert.* Geobios International, Jodhpur, India.

Shankar, V. 1980. Distribution of Khejri (*Prosopis cineraria* MacBride) in western Rajasthan. In: *Khejri (Prosopis cineraria) in the Indian Desert - Its Role in Agroforestry* (Mann, H.S. and S.K. Saxena, eds). CAZRI Monograph No. 11, CAZRI, Jodhpur, pp. 11-19.

Shukla, J.K. 2001. Ecological evaluation and adaptations in some plants of Indian desert. Ph.D. Thesis, J.N.Vyas University, Jodhpur, India.

Shukla, J.K., Chauhan, A., Kasera, P.K., Mohammed, S. and Chawan, D.D. 2000a. Seed variability in *Prosopis cineraria* collected from different localities of Indian desert. *Sci. & Cult.* **66**: 163-164.

Shukla, J.K., Kasera, P.K. and Chawan, D.D. 2000b. Effect of different treatments on growth, survival value and harvest index of *Prosopis cineraria* (Linn.) Druce. *Haryana agric. Univ. J. Res.* **30**: 65-70.

Singh, C.B., Idnani, L.K. and Singh, H.K. 2000. Response of irrigation and fertilizer levels on growth and biomass accumulation of *Eucalyptus* and *Leucaena* under dryland conditions. *Indian J. Soil Cons.* **28**: 58-60.

Singh, K.S. and Lal, P. 1969. Effect of Khejri (*Prosopis spicigera* Linn.) and Babool (*Acacia arabica*) a trees on the soil fertility and profile characteristic. *Annals Arid Zone* **8**: 33-36.

Singh, V. and Pandey, R.P. 1998. *Ethnobotany of Rajasthan, India,* (eds.). Scientific Publishers (India), Jodhpur.

Solanki, K.R., Kackar, N.L. and Jindal, S.K. 1986. Air layering in *Prosopis cineraria* (L.) MacBride. *Indian Forester* **112**: 202-207.

Tiwari, K.P. and Pandey, R.K. 2000. Development of agro-techniques for cultivation of medicinal plants used in Ayurveda, Unani, Siddha and Homeopathy. *III Annual Progress Report,* ISMH Project. State Forest Research Institute, Jabalpur, pp. 46.

Ukani, M.D., Limbani, N.B. and Mehta, N.K. 2000. A review on the Ayurvedic herb *Prosopis cineraria* (L.) Druce. *Ancient Science Life* **20**: 58-70.

Chapter 10

Studies on Phloem of Some Common Ferns of Rajasthan

B.L. Yadav, M.K. Tripathi and K.D. Chawla*

INTRODUCTION

No enough work has been done so far on the phloem of pteridophytes. Lamoureux (1961) studied comparative structure of phloem in vascular cryptogams and indicated the possible occurrence of sieve-tube members in *Marsilea quadrifolia.* Agashe (1968) noted the presence of sieve tubes in *Equistum giganteum* and *E.arvense* and concluded that because of this the phloem is of fairly advanced type among pteridophytes. Kruatrachue & Evert (1974) observed the presence of pores only at end walls of the sieve element in *Isoetes muricata.* Shah & Fotedar (1974a) described the structure of phloem in the rachis of *Pteris longifolia* and stem of *Cyathea gigantia* and *Blechnum orientale.* Shah *et al* (1976) recorded the occurrence of branched sieve elements in rachis and stem/rhizome of several Indian ferns. They also recorded the presence of projections (with sieve areas) on the lateral walls of the sieve elements. Shah & Nair (1976) studied uncommon wall thickenings in sieve cells of *Pteris wallichiana.* Nair (1978) also studied phloem structure in some ferns.

Loyal & Maheshwari (1979) described the presence of sieve elements with transitional stages between typical sieve cells and sieve tube members in *Marsilea*. Fotedar & Bir (1980) worked upon the structure of phloem in *Ophioglossum* and *Botrychium* from India. Describing the structure of phloem in Ophioglossaceae, Fotedar & Bir (1981) pointed out that the sieve cells are elongated and enucleated containing at maturity proteinaceous refractive spherules and their walls are sculptured with sieve areas of different patterns. Some pores appear scattered without any configuration of a sieve area. Absence of callose in sieve cells is reported and they showed that there is no correlation between the sieve cell length and width. The length of end walls of the same sized sieve elements may have

Department of Botany, M.L.V. Government College, Bhilwara – 311 001

*Department of Biology, Jai Hind College, Mumbai - 400 020

variable length inclination. Malik & Bhardwaja (1986) studied the structure and ontogeny of phloem in the rachis of *Athyrium falcatum* from Mt. Abu, Rajasthan. A view regarding the sieve area specialization at the end walls of fern sieve elements was put fourth by Fotedar (1989).

Nair (1989) studied the ontogeny of the sieve cells in nine pteridophytes. Sieve cells formed earlier are narrow and thick walled then those formed later and seen towards the periphery of the vascular bundle. Fotedar (1995) recorded that mature sieve cells are elongated, enucleate with a thin parietal layer of cytoplasm containing protein spherules.

A thorough survey of literature indicates that this aspect of Rajasthan ferns has largely been ignored, therefore, an attempt has been made to investigate phloem in ferns of this region and the results are being presented in this paper.

MATERIAL AND METHODS

For the present study species of *Adiantum* and *Cheilanthes* commonly occurring and widely distributed in Rajasthan have been taken up. For phloem morphology, maceration of different parts (root, rhizome and rachis) was done to obtain entire and isolated sieve cells using Schultze's fluid. After maceration, material was washed with 5% Sodium bicarbonate ($NaHCO_3$) solution followed by washing in running water for about 6 – 8 hours and stained with Toluidine blue (Feder & O'Brien, 1968) before mounting in glycerine jelly. The cellulosic nature of sieve cell wall thickening was confirmed by I_2KI-H_2So_4 staining (Johanson, 1940).

OBSERVATIONS

Organ-wise description of sieve cell morphology of each of the taxa taken up for present investigation is given below. The parameters of sieve cell morphology are presented in Table 1-3.

TABLE 1

Morphology of sieve cells of some ferns of Rajasthan

Root

S.No.	Species	Length (μm)		Breadth (μm)	
		Range	Average	Range	Average
1.	*Adiantum capillus-veneris*	400-600	460±61	18-30	25±4
2.	*A.incisum*	425-617	566±18	11-18	14±2
3.	*A.philippense*	400-925	615±105	12-22	16±4
4.	*Cheilanthes bicolor*	325-925	660±114	11-20	14±3

ROOT (TABLE 1):

(a) *Adiantum* (Fig 1 B,D,E) : Mature sieve cell in root of *Adiantum incisum, A.philippense* and *A.capillus-veneris* is elongated with oblique to very oblique end walls. Walls of developing sieve cells are thick due to deposition of cellulose and contain elongated nuclei, which degenerate during course of maturation.

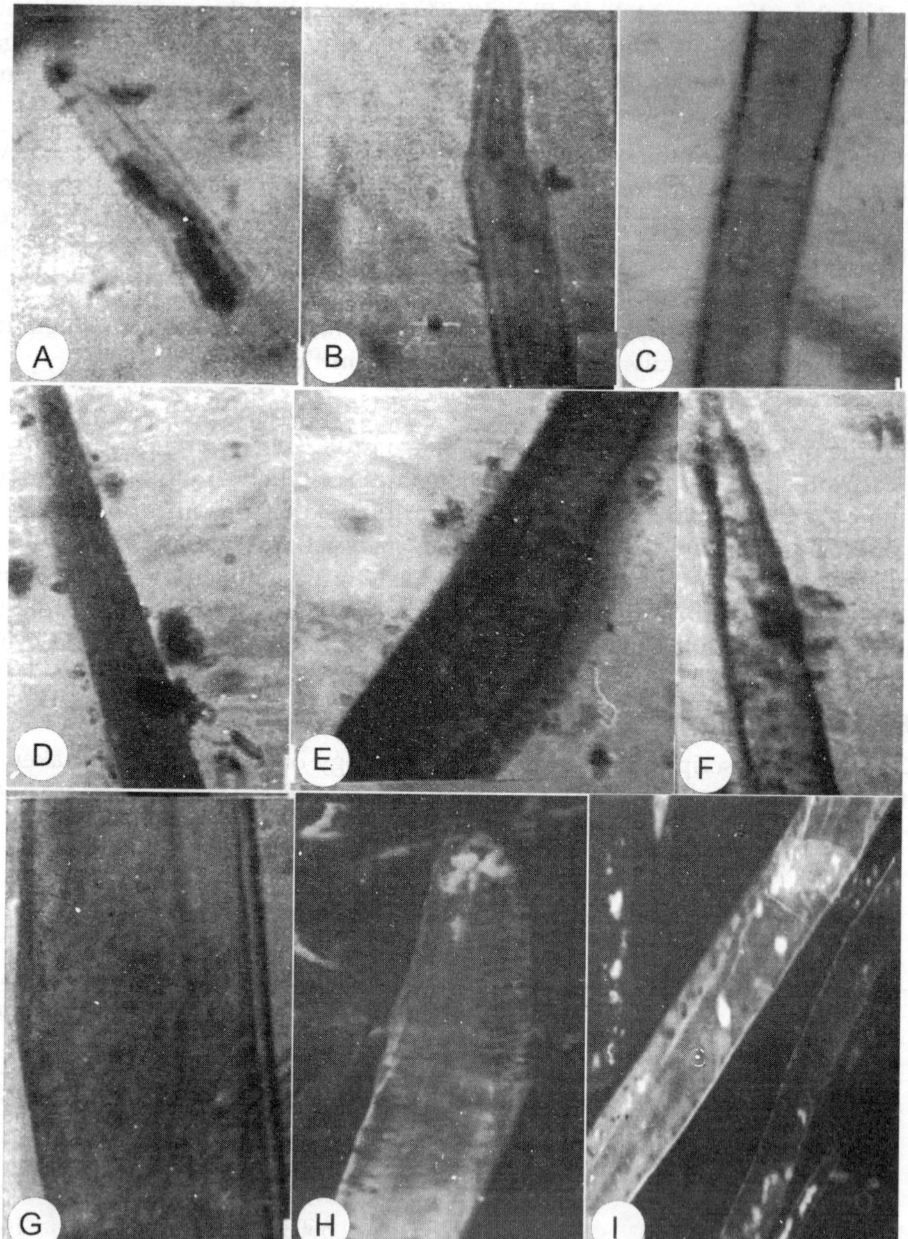

Fig. 1 (A-I) : *Adiantum capillus-veneris* (A-B) Rhizome A. Sieve cell with shrunken cytoplasm, Root. B. Sieve cell with oblique end wall, *A. incisum* (C-D) Rachis C. Cellulosic thickening on developing sieve cell. Root D. Lateral wall with pores in a leaner row pattern. *A. philippense* (E) Root E. Callose deposition on lateral walls, *Cheilanthes bicolor* (F-I) : Rachis F. Sieve areas on oblique end walls, Rhizome G. Cellulosic thickening on lateral walls. I. Sieve cell with primordial pits. Root H. Sieve cell with reticulate sieve areas at the end walls.

Average length of sieve cells was found to be 460 ± 61 µm in *A.capillus-veneris* while in *A.incisum* it is 566 ± 182 µm and 615 ± 105 µm in *A.philippense.* Ranges of sieve cell length in *A.capillus-veneris* was observed to be 400 – 600 µm while it is 425 – 617 µm in *A.incisum* and 400 – 925 µm in *A.philippense.*

Average breadth of the sieve cells was found to be 25 ± 4 µm in *A.capillus-veneris* while in *A.incisum* it is 14 ± 2 µm and 16 ± 4 µm in *A.philippense.*

Range of sieve cell breadth in *A.capillus-veneris* was found to be 18 – 30 µm while it is 11 – 18 µm in *A.incisum* and 12 – 22 µm in *A.philippense.*

Thus, in root mature sieve cells were longer in *A.philippense* and broader and shorter sieve cells were recorded in *A.capillus-veneris.* The size of sieve cells in root of *A.incisum* is ranging between these two. Sieve cells with transverse end walls have also been observed in root of *A.incisum.*

Sieve areas on the lateral wall are frequently present in a linear arrangement. These are restricted to the end walls in a reticulate pattern and on the middle portion these are infrequent.

(b) *Cheilanthes bicolor* (Fig. 1H) : In *Cheilanthes bicolor* the mature sieve cells are elongated with oblique end walls. The average length was recorded to be 660 ± 114 µm with the average breadth of 14 ± 3 µm. Range of sieve cell length was found to be 325–925 µm and the range in breadth was recorded to be 11 – 20 µm.

The sieve cells are longest and narrower in *C. bicolor* among the investigated leptosporangiate ferns.

Sieve cells have primordial pits throughout their lateral walls. Sieve areas are frequently present on lateral walls of mature cells in a linear arrangement. Sieve cell with transverse end walls have also been observed.

Rhizome (Table 2)

(a) *Adiantum* (Fig 1 A) : Mature sieve cells in rhizome of three species of *Adiantum* occurring in Rajasthan are elongated with thick lateral walls due to deposition of cellulose. In some sieve cells heavy deposition of callose is observed. End walls of sieve cells are oblique. Perforation on end walls is also observed. Sieve areas are frequent and are arranged in reticulate pattern. Large size pores on lateral walls in a mature sieve cell are observed.

Lobing of nucleus is observed in sieve cells of *A.capillus-veneris.* Mature sieve cells with transverse end walls were also observed in *A.capillus-veneris* and branched sieve cells were seen in *A.incisum.* Sieve cells with varied shapes were observed in rhizome of all the three species.

Maximum range in the length of sieve cells was observed in *A.capillus-veneris* (125-350 µm) followed by *A.incisum* (137 - 205 µm) and the minimum range was recorded in *A.philippense* (112 – 175 µm).

Sieve cells with highest value of average length was found in *A.capillus-veneris* (222 ± 72 µm) and the lowest in *A.philippense* (140 ± 21 µm) while *A.incisum* with 179 ± 23 µm value falls in between.

TABLE 2

Morphology of sieve cells of some ferns of Rajasthan

Rhizome

S.No.	Species	Length (μm)		Breadth (μm)	
		Range	Average	Range	Average
1.	*Adiantum capillus-veneris*	125-350	222±72	35-62	50±12
2.	*A.incisum*	137-205	179±23	10-19	15±4
3.	*A.philippense*	112-175	140±21	15-25	20±3
4.	*Cheilanthes bicolor*	137-400	252±95	15-31	21±6

(b) *Cheilanthes bicolor* (Fig 1G, I) : Mature sieve cells in rhizome of *Cheilanthes bicolor* are elongated with oblique to very oblique end walls. Lateral walls of sieve cells were found to be thicker than the parenchymatous cells. Sieve areas are frequent and are arranged in a reticulate pattern. Alternate arrangement of sieve areas was rarely observed. Pores on lateral walls of mature sieve cells were frequent, sieve cells with varied shapes are common.

Range of sieve cell length was observed to be 137 – 400 μm. Average length of mature sieve cell was found to be 252 ± 95 μm. The range of sieve cell breadth was recorded to be 15 – 31 μm and the average breadth of the mature sieve cell was observed to be 21 ± 6 μm.

RACHIS (TABLE 3)

(a) *Adiantum* (Fig 1 C, 2A-C,E-F,H,J) : In all the three species of *Adiantum* occurring in Rajasthan mature sieve cells are profusely elongated, enucleated with oblique to very oblique end walls. It possesses thick wall and parietal cytoplasm. Sieve areas occur frequently in the middle segment of the lateral walls of sieve cell in these species. Primordial pits occur frequently on lateral walls of young sieve cells. Sieve areas have been observed in a reticulate pattern. These remain filled with pores. Lateral wall also contain pores in a linear row pattern. Deposition of callose is observed on end walls of sieve cells.

TABLE 3

Morphology of sieve cells of some ferns of Rajasthan

Rachis

S.No.	Species	Length (μm)		Breadth (μm)	
		Range	Average	Range	Average
1.	*Adiantum capillus-veneris*	350-800	590±180	7-26	14±6
2.	*A.incisum*	686-1062	892±141	11-18	13±6
3.	*A.philippense*	525-1125	742±215	7-18	13±4
4.	*Cheilanthes bicolor*	425-700	580±102	21-31	24±4

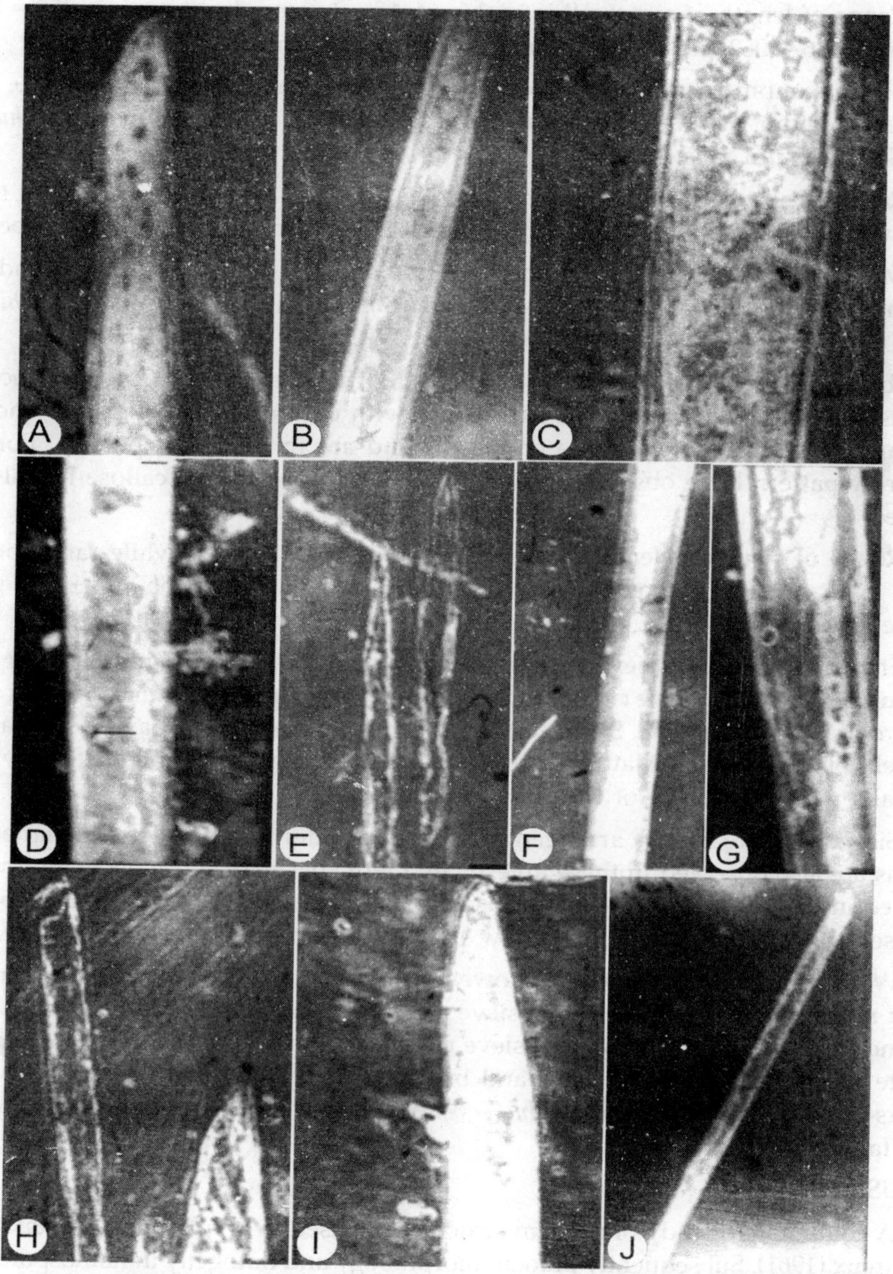

Fig. 2 (A – J) Rachis : *Adiantum capillus-veneris* (A-C,F) : A. Sieve cell with scattered sieve areas. B. Sieve cell with short oblique end walls. C. Sieve cell with alternate sieve areas F. Sieve cell with alternate sieve areas, *A.incisum* (J) : J. Primordial pits on lateral wall. *A.philippense* (E-H) : E. Callose deposition on lateral walls. H. Sieve areas developing from primordial pits. *Cheilanthes bicolor* (D,G,I) : D. Alternate sieve areas on lateral walls. G. Sieve areas on lateral walls. I. Alternate sieve areas on lateral walls.

Range of length of sieve cell in rachis of *Adiantum capillus-veneris* is 350 – 800 µm while it is 686- 1062 µm in *A.incisum* and 525 – 1125 µm in *A.philippense*

The maximum average length of sieve cell was recorded in *A.incisum* (892 ± 141µm) and the minimum was found in *A.capillus – veneris* (590 ± 180 µm) while *A.philippense* with 742 ± 215 µm falls in between.

Range of sieve cell breadth in petiole of *A.capillus – veneris* was observed to be 7–26 µm while in *A.incisum* and *A.philippense* it is 11 –18 µm and 7 – 18 µm respectively.

Maximum average breadth is observed in *A.capillus – veneris* (14 ± 6 µm) and almost equal average breath value was observed in *A.incisum* (13 ± 6 µm) and *A.philippense* (13 ± 4 µm.)

(b) *Cheilathes bicolor* (Fig 1 F, 2 D,G,I) : Mature sieve cell in *C. farinosa* occurring in Rajasthan was quite elongated, enucleated with oblique to very oblique end walls. Sieve areas were restricted to the end walls and are in reticulate pattern. Pores in a linear row pattern have observed on lateral walls. Deposition of callose has also been observed.

Range of sieve cell length was found to be 425 – 700 µm while range of breath was 21 – 31 µm. The average sieve cell length was recorded to be 580 ± 102 µm and average breadth 24 ± 4 µm.

Structural details of sieve cells in the selected fern species manifest an organ-wise differentiation pattern. Sieve cells of rachis are quite elongated but thin compared to those of the root and the rhizome. Sieve cells in the rhizome are very broad in comparison to sieve cells of other organs. Mature sieve cells in each organ possess oblique to very oblique end walls with deposition of callose on lateral walls.

Sieve areas are mostly arranged in reticulate pattern while alternate and scattered patterns are also observed in these investigated fern species. Callose deposition was also observed in the sieve areas. Lateral walls of sieve cells in all organs have been observed to possess pores arranged in linear row pattern or scattered on whole surface.

Walls of mature sieve cell are however, thicker than walls of phloem parechyma cells. It may be emphasized here that sieve areas occur only on a wall between two sieve cells and never on a wall between a sieve cell and parenchyma cell. These observations indicate that sieve cells are shorter and broader in taxa occurring in moist or aquatic habitats as found in *Adiantum capillus – veneris* which occurs along water channels or water falls.

DISCUSSION

A comparative study of phloem structure in vascular cryptogams was taken up by Lamoureux (1961). Subsequently phloem morphology and ontogeny in homosporous ferns has received much attention during the recent past (Shah & Fotedar, 1974; Shah & Nair, 1976; Nair, 1978; Fotedar & Bir, 1980; Malik & Bhardwaja, 1986). Structural details of phloem of the land and water forms of some species of *Marsilea* have been studied by D'souza (1990).

During present investigations, structural details of sieve cells in the selected fern species indicate that sieve cells of rachis are quite elongated but thin compared to those of the root and the rhizome. Sieve cells in the rhizome are very broad in comparison to sieve cells of other organs. Sieve areas are mostly arranged in reticulate pattern while alternate and scattered patterns are also observed in these investigated fern species. Callose deposition was observed in the sieve areas. Lateral walls of sieve cells in all organs have been observed to possess pores arranged in linear row pattern or scattered on whole surface.

Cellulosic wall thickening was present in young sieve cells of these species but in mature sieve cells these walls become comparatively thinner. Such thickening of sieve cell walls has also been recorded by Burr & Evert (1973) in *Selaginella krausseana* and Krutrachue & Evert (1974) in *Isoetes muricata.* Walls of mature sieve cells are however thicker than walls of phloem parenchyma.

Mature sieve cells in each organ possess oblique to very oblique end walls with deposition of callose on lateral walls as also on end walls. Heavy callose deposition has been observed on end walls and lateral walls of the mature sieve cells of rhizome. Kruatrachue & Evert(1974) and Warmbordt & Evert [1974] however, reported the absence of callose in such pteridophytes as *Isoetes municata and Lycopodium lucipulum.* According to Lamoureux (1961) deposition of callose is related with seasonal changes.

REFERENCES

Agashe, S.N. 1968. Phelom studies in Pteridophytes Part I *Equisetum. Am. Fern J.* 58: 74-77.

Burr, F.A. & Evert, R.F. 1973. Some aspects of sieve element structure and dovelopment in *Selaginella kraussiana Protoplasm,* 78: 81-97.

D'Souza, M.I.C. 1990. Amphibious habit with particular reference to land and water forms in *Marsilea* L. Ph. D. Thesis, University of Rajasthan, Jaipur.

Feder, N. & O'Brien, T.P. 1968. Plant microtechniques. Some principles and new methods. *Amer. J. Bot.* 55: 123-143.

Fotedar, R.L. 1989. Probable trends of sieve area specialization in ferns. *Phytomorphology* 39 (2,3): PP 169-172.

Foteder, R.L. 1995.Sieve cell ontogeny in *Helminthostachys zeylanica* and *Angiopteris evecta. Indian fern J.* 12: 79-87

Fotedar, R.L. & Bir, S.S. 1980. Sturucture of phloem in two members of Ophioglossaceae from India. *Phytomorphology* 30(2,3): 164-172.

Fotedar, R.L. & Bir, S.S. 1981. Structure of phloem in two members of Oplioglossaceae from India. *Phytomorphology* 30: 164-172.

Kruatrachue, M. & Evert, R.F. 1974. Shucture and development of sieve element in the leaf of *Isoetes muricata. Amer. J. Bot.* 61: 253-266.

Lamoureux, C.H. 1961. Comparative studies on phloem of vascular cryptogames. Diss. Univ. Calif. Davis.

Loyal, D.S. & Maheshwari, A.K. 1979. Sieve cells vs. sieve tube. members in *Marsilea*, a re-evaluation. *Phytomorphology* 29: 68 - 70.

Malik, S.K. & Bhardwaja, T.N. 1986. Petiolar anatomy and ontogeny of pheloem in rachis of *Athyrium. Indian fern J.* 3: 35-99.

Nair, M.N.B. 1978. Structure and development of Phloem in ferns. Ph. D. thesis. Sardar Patel Univ. Vallabh Vidyanagar, Gujrat, India.

Nair, M.N.B. 1989. Ontogeny of sieve cells in some Pteridophytes. Phytomorphology 39: 173-180.

Shah, J.J. & Fotedar, R.L. 1974a. Structure and development of phloem in the rachis of *Pteris longifolia*. *Phytomorphology* 24: 107-113.

Shah, J.J., Fotedar, R.L., Nair, M.N.B. & Kothari, I.L. 1976. Occurrence of branched sieve elements. *Acta Botanica Indica* 4: 53-55.

Shah, J.J. & Nair, M.N.B. 1976. Uncommon wall thickenings in the sieve cells of *Pteris wallichiana*, *Amer. Fern. J.* 66 (1): 19-20.

Warm brodt, R.D. & Evert, R.F. 1974. Structure and development of sieve element in the stem of *Lycopodium lucidulum*. *Am. J. Bot.* 61: 267-277.

❑❑❑

Chapter 11

Effect of Human Interferences on Orchid Diversity

Purshotam Kaushik and Pallavi Pal

Orchids are the most beautiful flowering plants exhibit an incredible range of diversity in size, shape and colour of their flowers. They are most pampered of the plants and occupy top position among all the flowering plants valued for cut flower production and as potted plants, and thus orchids have a huge horticultural market worth 100 million dollar annually in the U.S. alone (Griesbach, 2002). Several orchids have therapeutic importance and find extensive use in Ayurvedic local medicines (Kaushik, 1983; Handa, 1986) because they are rich in flavonoids, glycosides, carbohydrates and other phytochemical contents (Husen and Rahman, 2003). The orchids are also known to have alkaloids and possess antibacterial potential (Ghanaksha and Kaushik, 2007).The famous and precious Chinese drug 'Shih-leu' which consisted of dried fresh stems of certain species of *Dendrobium* was widely used as a strengthening agent and tonic was also considered to prolong life, to relieve pain, as anti-inflammatory, to stop vomiting and excessive perspiration.

Land clearance, industrialization, deforestation through burning of forest trees and falling of trees for timber has been the major cause for the depletion of Indian orchid wealth and other human activities having an adverse impact on the wild orchid population and associated fungi. A large number of orchid species which were present in plenty in Indian forests are now at the verge of extinction and some of them have become so rare that a large number of botanical teams were unable to trace them. To cite an example *Paphiopedilum druryi* which was once found in plenty in Agastaya hills in South India is now difficult to locate. Steps should therefore, be taken to conserve this natural wealth, obviously, the protection of orchid population and orchid associated fungi is important

Department of Botany and Microbiology, Gurukul Kangri University, Hardwar - 249 404 in maintaining global biodiversity and it also has implications for overall ecosystem health.

DISTRIBUTION AND HABITAT

Taxonomically, they represent the most highly evolved family among monocotyledons with 600-800 genera and 25,000-30,000 species. It is estimated that about 1,600 species including 151 genera of orchids are found in our country in the Himalayas as their main home and others scattered in eastern and western ghats. The regional distribution has been estimated to be as: peninsular India, 250 species, north-east region 876 species, the eastern Himalayas 612 species, and north-western Himalayas 255 species. The north-eastern region is considered as the most important hot spot of orchid biodiversity of the Indian subcontinent due to the favourable climatic conditions such as humidity, low temperature accompanied by good rainfall.

Members of the Orchidaceae are perennial herbs widely distributed in both hemispheres. The family is cosmopolitan and believed to be undergoing rapid diversification and speciation with members found in a vast range of habitats. Mainly orchids are found in three habitats firstly terrestrial or soil dwelling, epiphytic which grows on branches and tree trunks and which grow on rock surfaces are termed as lithophytic orchids. Terrestrial orchid penetrating the tissues in the surface substrate of the ground such as soil or organic floor. Epiphytic orchid use trees as a platform upon which they grow with in the canopy, they use trees as a platform only and they do not obtain their moisture or nutrients from the plant whereas lithophytes grow on the surface of rocks or in the rock cervices.

INDIAN ORCHIDS

Some of the Indian orchid species which have been regarded to be of high ornamental value are : *Aerides crispum, Aerides fieldingii, Aerides multiflorum, Aerides odoratum, Anoectochilus roxburghii, Arachnis clarkei, Arundina graminifolio, Bulbophyllum leopardinum, Calanthe masuca, Coelogyne elatn, Coelogyne devonianum, Cymbidium pendulum, Cymbidium longifolium, Cymbidium munronianum, Dendrobium aggregatum, Dendrobium aphyllum, Dendrobium fimbriatum, Dendrobium jenkinsii, Dendrobium moschatum, Dendrobium nobile, Paphiopedilum faireanum, Paphiopedilum venstum, Paphiopedilum hirsutissium, Paphiopedilum insigne, Phaius wallichii, Pleione praecox, Rhynchostylis retusa, Thunia alba, Vanda cristata, Vanda coerulea and Vanda coerulescens.*

LIST OF SPECIES UNDER THREAT

The conservation status of the 372 taxa of orchids native to the Australian State of Victoria was assessed using the IUCN 2001 Red List categories and criteria. The assessment resulted in the following findings: extinct, 11 taxa (10 global, 1 regional); critically endangered, 79 taxa (70 global, 9 regional); endangered, 26 taxa (12 global, 14 regional); vulnerable, 92 taxa (38 global, 54 regional); near Threatened, 8 taxa (1 global, 7 regional); data deficient, 24 taxa (15 global, 9 regional); and least concern, 132 taxa (Backhouse and Cameron, 2005). Therefore, there is a need to investigate the orchid diversity of unexplored and under explored areas.

TABLE 1

Global conservation status of some orchids

S. No.	Orchid Species	Category
1.	*Taeniophyllum scaberulum* Hook. f.	Extinct
2.	*Vanda thwaitesii* Hook. f.	Extinct
3.	*Vanda corulea*	Rare and endangered
4.	*Platanthera praeclara*	Rare and endangered
5.	*Renanthera imschootiana*	Rare and endangered
6.	*Herminium lancenum*	Endangered
7.	*Dactylorhiza hatagirea* (D.Don) Soo	Endangered
8.	*Phalaenopsis lindenii*	Endangered
9.	*Laelia speciosa*	Endemic & endangered
10.	*Caladenia fragrantissima*	Endangered
11.	*Galeola falconeri* Hook. f.	Endangered
12.	*Thelymitra malvina*	Vulnerable
13.	*Calypso bulbosa*	Vulnerable
14.	*Thelymitra circumsepta*	Vulnerable
15.	*Caladenia robinsonii*	Critically endangered
16.	*Paphiopedilum praeclara*	Critically endangered

LOSS OF ORCHID DIVERSITY: A CASE STUDY

Several studies have pointed out the significance of habitat destruction and overexploitation in the endangerment of wild orchids from different parts of the world (Alphonso, Borromeo, Pradhan and Rogaly, 1975).

A total of 960 species of orchids belonging to 158 genera have been recorded from the Indian Himalaya region. *Bulbophyllum* (86 spp.), *Dendrobium* (82 spp.), *Eria* (45 spp.), *Liparis* and *Oberonia* (34 spp. each), *Habenaria* (30 spp.), *Cymbidium* (27 spp.), *Peristylus* (25 spp.) and *Calanthe* (23 spp.) were the species rich genera. The Sikkim and Darjeeling hills of central Himalaya and Arunachal Pradesh of eastern Himalaya harbour about 60% of the total orchids recorded from the Indian Himalayan Region. The tropical and subtropical zones showed the maximum richness of orchids (95.10%). Nearly 59.58% species were native to the Himalayan region and neighbouring countries together, 30.63% species endemic and 16.46% species near endemic to the Indian Himalayan region. Orissa is also rich in orchid flora 130 species of orchids are found in Orissa.

The high anthropogenic activities particularly in the tropical, subtropical and temperate zones have caused habitat degradation and population depletion of the orchid species. Invasion by exotic species in these zones has also contributed to the population depletion of the orchid species. Besides this climatic variation has also contributed to the habitat alterations and loss of orchid diversity.

Balakrishnan (1981) has provided a graphic description of decimation of the vegetation of the north-east region. In 1851 J.D. Hooker, on his expedition to Jaintia hills had collected seven head loads of live *Vanda coerulea* (Blue Vanda orchid) plants for cultivation in England. Balakrishnan states that after 100 years, during various field trips stretching from 1965 to 1970, he could hardly spot a dozen plants even in remote forest areas of Nertiang where Hooker had made his collections. This is an indication of the rate at which primary forests and orchid wealth are being destroyed.

Dactylorhiza hatagirea (D. Don) Soo, earlier known *Orchis latifolia* Hook (Vij *et al.* 1992), it is a terrestrial orchid and commonly known as 'Salampanja' and 'Hatajari' in Garhwal Himalaya is of medicinal value. The tubers of this species, commonly sold as salampanja are known to yield a high quality 'Salep' which is extensively used in local medicine as nervine tonic for its astringent and aphrodisiac properties (Vij *et al.*, 1992). It has been categorized as critically endangered and rare (Samant *et al.*, 2001) and listed under appendix II of CITES (Convention on International Treaty of Endangered Species of Flora and Fauna).

The various changes in the Himalayan forests are appearing in their structure, density and composition due to global warming, uncontrolled lopping and felling of trees for fuel wood, fodder and grazing (Gaur 1982, Bargali *et al.*, 1998; Kumar *et al.*, 2004). Changes in climate, exploitation of several plants for medicine and grazing pressure in alpine region of Garhwal Himalaya have led to drastic changes in vegetation composition and population of species during last few decades (Nautiyal *et al.*, 2004). Therefore it is a need to promote cultivation, propagation and conservation of this species. Using *in-situ* as well as *ex-situ* conservation efforts, we can propagate and conserve this species and an of other ecologically as well as economically important plants of high altitudes.

According to Das and Khumbongmayum (2006) Arunachal Pradesh is part of eastern Himalayan ranges and is known for its rich biological diversity and state is a proud of India. The extremely variable topography, altitude, climate, vegetation support high biological diversity favours the state to hold the top position for having maximum number of orchid species. About 550 species of orchids under 110 genera have been reported so far from Arunachal Pradesh, out of which 370 are epiphytes, 160 are terrestrial and 20 are saprophytes and 2 are lithophytes, surveyed the area during the July 2005 during the course of extensive exploration of plants in and around the Ziro plateau (26°50¢–28°21¢N lat. and 92°40¢–94°21¢E long.) in lower Subansiri district of Arunachal Pradesh; a giant saprophytic orchid was collected. It was identified as *Galeola falconeri* Hook. f. is an endangered species and rare in occurrence.

Galeola falconeri is sensitive to the environment that requires a specific niche in the environment, but the recent survey revealed that this species could not be found elsewhere in and around the forest. Destruction of the protective canopy of the forests due to felling of large tree species for agriculture or horticultural activities and other developmental purposes have adversely affected the physicochemical variables of the soil and microclimatic conditions prevailing in the forest. Such activities have changed the mosaic of the natural habitat of many valuable orchid species. Conservation and management of orchids require a proper understanding of the reproductive biology and concerted efforts to preserve this valuable orchid in their natural habitats.

Xavier and Senthilkumar (2007) carried out intensive field surveys in hilly tracts of the eastern ghats of Tamil Nadu these were the potential site of orchids having luxuriant and gregarious orchid vegetation in different areas. Some medicinally important orchids of area are *Acampe praemorsa, Anoectochilus elatus, Bulbophyllum fischeri, Cymbidium aloifolium, Malaxis acuminata, Rhynchostylis retusa* and *Dendrobium nanum* etc. However, the present status of these orchids is that many of them are on the verge of extinction due to over exploitation by human beings for their own benefit. The disappearance of orchids is also due to lack of proper documentation of economic potential and biological functioning of orchids, besides unawareness among local inhabitants.

Ramakantha (2009) recorded that Northeast India sustains the highest concentration with about 700 species. As many as 34 species of orchids from north-east India are listed among the threatened plants of India. The following are some ornamental orchids of the region: *Paphiopedilum fairieanum, Paphiopedilum insigne, Paphiopedilum villosum, Paphiopedilum spicerianum, Paphiopedilum hirsutissimum, Paphiopedilum venestum, Anoectochilus sikkimensis, Vanda coerulea, Vanda teres, Renanthera imschootiana, Pleione maculata, Cymbidium eburneum, Dendrobium hookerianum, Dendrobium densiflorum, Dendrobium devonianum, Dendrobium thyrsiflorum* and *Thunia marshalliana*. Many species of orchids are of medicinal importance. For example, the fresh as well as dried stem of the orchid *Dendrobium nobile* is used in the preparation of the Chinese drug 'Shih-leu' used as an aphrodisiac, analgesic and for longevity.

The primary vegetation in extensive areas of the north-east India has been disturbed and modified and in some places destroyed by seismic activities, frequent landslides and resultant soil erosion. In all, at present more than 700 species of plants from the north-east India is facing the threat of survival in the wild.

Pattanaiki *et al.* 2009 emphasized the need of conservation in Araku valley in Andhra Pradesh. Due to high elevation, average rainfall of 1700 mm, and mixed deciduous forests with a luxuriant growth of orchids, ferns and epiphytes. The forests to a great extent seem to be secondary in nature, probably due to extensive shifting cultivation practiced by local and tribal people. The proposed bauxite mining activity by Andhra Pradesh Mineral Development Corporation Limited in these areas would not only wipe out the virgin forests, but also destroy the habitats of several endangered flora.

A conservation awareness programme would help make the villagers, local forest guards and NGOs participating in conservation efforts understand the importance of endangered and threatened flora-fauna as well as the forests surrounding them. Further research work is needed to map and monitor the habitats of threatened species. Very-high resolution satellite datasets (Cartosat-1, Cartosat-2 and Resourcesat-1 LISS IV) should be a requisite which can help the conservationists and decision makers to focus on the distribution of species while designing future conservation strategies.

The Frankston Spider-orchid *Caladenia robinsonii* is one of the most threatened orchids in Australia. It is included in the ANZECC List of Threatened Australian Flora as endangered, is listed as a threatened taxon under the Victorian *Flora and Fauna Guarantee Act* 1988, and is considered endangered by Carr (1991), Backhouse and Jeanes (1995); Briggs and Leigh (1995). Under current IUCN (International Union for Conservation of

Nature and Natural Resources) Red Book threat categories and criteria (1994), the Frankston Spider-orchid is critically endangered.

Habitat destruction for urban development has been the major cause of the decline to virtual extinction of the Frankston Spider-orchid. Current major threats include weed invasion, habitat degradation, altered fire regimes, disturbance and possible illegal collection (Backhouse *et al.*, 1999).

The Orchidaceae family in Nepal ranks third in size in the angiosperm group. This family consists of 97 genera and 363 species making one of the largest groups of endemic plant species. Churiya stands out as it is fragile and has dynamic characteristic landscape. It is identified as a potential area in terms of natural and forest resources. Convention on International Treaty of Endangered Species of flora and fauna (CITES) was established in 1973 and Nepal joined as a state party in 1975. CITES has three different categories of ranking species, critical endangered, endangered and threatened. It has incorporated orchid family as endangered species.

Huge amount of orchids are collected illegally and in unscientific way. Orchid species are exploited for medicinal purpose from tropical and sub-tropical regions (Bailes, 1985). Threats to conservation of orchids in Nepal are associated with habitat loss, forest destruction, and degradation and over exploitation. Unimaginable scale of orchids along with its plant is harvested from Basamadi, Sannanitar, and Makawanpur district everyday. Makawanpur, Lothar–Chitwan, also bear similar brunt. At the rate of which it is harvested, extinction of these species from the churiya range is not too far. Average collection reaches 35 - 50 quintals a day. This illicit scheme continues from December to April but during January, February and March–it is over-harvested because orchids flower during that period. Deducting and multiplying the number of harvesting days, we assume five thousands quintals of orchids are harvested from the churiya forest, which is not a good omen for Nepal's orchids. Orchids are used as weeds, compost, scattered, bedding and cattle feed. The poor collect it and are meagerly compensated with little money.

FACTORS RESPONSIBLE FOR LOSS OF ORCHID DIVERSITY

Rapid orchid diversity loss world wide could be attributed to the following reasons:

1. Human Interferences

A number of orchid species have lost their habitat because of increased human interference like construction of dams, roads, railway tracts or bridges across natural ecosystems, mining activity or industries. Habitat fragmentation due to deforestation or any other biotic stress has caused much harm to precious biodiversity. According to the Forest Survey of India in 2000, deforestation rate is increasing due to burgeoning population and this has destroyed the natural homes of species. However, the forest cover is rapidly disappearing from the entire region. There has been a decrease of about 1800 sq.km. in the forest cover between 1991 and 1999. More worrisome still is the fact that the quality of the forest is also deteriorating, with the dense forests (canopy closure of 40% or more) becoming degraded into open forest or scrub. Though there is a succession of several edaphic formations, a vast area of land has already been transformed into barren and unproductive wastelands. This being the case, the statistics of more than 64 % of the total geographic area in this region under forest cover could be misleading.

For example, though the forest cover in Manipur extends to 78% of the total geographic area, only 22% of forest area is under dense forest cover and the rest has been converted to open forests. The primary vegetation in extensive areas of the India has been disturbed and modified and in some places destroyed by seismic activities, frequent landslides and resultant soil erosion. While these natural causes have contributed only marginally to the change in vegetation type, it is the activity of man that has led to the irreversible transformation in the landscapes and has resulted in colossal loss of biodiversity in the entire region. Human influences have pushed many species to the brink of extinction and have caused havoc to natural fragile ecosystems. Such devastations to natural ecosystems are witnessed almost everywhere in the region and is a cause of great concern.

2. Grazing

It is estimated that 60% of the domestic herbivore population graze in the forest. The grazing causes soil compaction and heavy damage to the forest plantations and natural regeneration. This far exceeds the carrying capacity of the forests and is one of the major factors for degradation of forests especially near habitations. Barren and unproductive cattle in spite of having uses like transport, manure, capital etc are generally viewed as a great drain on the resources.

This report provides a case study on the pressure of grazing in the Barsey Sanctuary of Sikkim and the efforts of the Forest Department in Sikkim in reducing the problem through the eco development committees. The Arunachal Pradesh SBSAP (State Biodiversity Strategy and Action Plan Reports) reports that biotic pressures are mostly in the form of grazing, trampling and browsing This is very prevalent in foot hill regions adjoining Assam where large herds of cattle intrude and damage new regenerations and palatable plants. In most of the young forest plantations this is a growing problem. In higher elevations similar damage are caused by sheep and yak. Apart from grazing or browsing, these animals are responsible for trampling and damage to young recruits to such disturbance under present socio-economic conditions is practically impossible.

3. Forest Fires

Forest fires are common and frequent affecting about 20% of the total forest area. Such fires at the end of winters are a rule rather than exception. As per assessment of FSI (1993), it is estimated that around 6.16% of the forests are heavy to moderately degraded by fire (SBSAP Tripura) in places, retrogression of vegetation has already set in. The main constraints in control of forest fires are lack of communication facility for early detection of forest fire and also the lack of fire fighting equipment. Forest fires sweep a large area of forest floor every year, affecting regeneration in natural forests and young crop in the plantations. Forest fires are reported from the pine forests of Arunachal Pradesh resin tappers who are contracted by the village councils often do not take precautions, subjecting the forests to accidental fires. Although fire is an ecological requirement for pine, accidental or deliberate fires causes immense damage to the forest ecosystems.

4. Illegal Extraction

As orchids are epiphytic plants and takes many years to grow and flower, there is a high possibility that they may never come back from the brink of extinction. To counter such flaw within the department, frequency of orientation programme on banned flora and fauna must be conducted by concerned line agencies. The customs must be given posters of species that are banned and manuals to be effective.

There must be an understanding amongst line agencies to be effective in implementing international conventions and a way sought to counter such activities.

5. Un-coordinated Infrastructure Development

Due to lack of deep environmental awareness in policy-making, many models of development are proving to be unsustainable, and the dimensions of such negative impacts are as yet unclear, as monitoring, baseline data, impact assessment and threat or risk assessment are poorly developed in the states. In Arunachal Pradesh, activities like road construction which are an essential component of developmental infrastructure have been at times the roads to destruction of biodiversity. It damages not only the area over which it passes but also causes land slides that destroy patches of forests. The construction of roads also leads to depletion of fire wood, poles and timber.

6. Lack of Awareness

Common people should be aware about the importance of orchids. A conservation awareness programme would help make the villagers, local forest guards and NGOs participating in conservation efforts understand the importance of endangered, threatened flora and fauna as well as the forests surrounding them. Awareness rising amongst those who influence development decisions, including local people, is therefore essential. This awareness raising is resource intensive and usually falls to the under resourced voluntary sector, so it cannot always be done successfully.

Above cited case studies clearly indicated that orchid diversity is on the verge of extinction and its time to conserve it otherwise we are bound to loose our gems for ever. The following are some suggestion and measures that we must taken to protect these gorgeous flowers.

PROTECTIVE MEASURES

Various factors like human settlement, road broadening, deforestation, grazing, lack of knowledge among local inhabitants etc are the major factors responsible for declining of orchid diversity. There is a need to check against all these activities so as to protect our natural gems.

- Regional surveys should be made to note up the data of the particular orchid species.
- Awareness should be created among common people.
- Wild orchid species should be protected, as they are more prone to deforestation.
- Beside this, Orchidarium can be considered as a source of material for scientific investigations, a wide range of educational programs, as well as an efficient tool in raising public awareness in issues related to conservation of rare orchid species.
- *In-vitro* propagation and tissue culture techniques are the some biotechnological tools which solve the problem to an extent.

- *Ex-situ* cultivation of orchids should be maintained.
- Government must take some measures to check the construction of roads and houses on the areas where plants having orchid habitats.

CONCLUSION

Therefore, for the sustenance of the biodiversity and smooth functioning of the natural cycles, the role of environmental education at the grass roots level is very important. Through this the demonic person can be transformed in divine person. Conservation of nature would only help to save the orchids. Conservation and management of orchids require a proper understanding of the reproductive biology and concerted efforts to preserve their habitats. Therefore, an extensive inventory of the species is required to understand their specific habitat, location, associated plants, population structure, etc. Scientific conservation measures are needed along with developmental activities so as not to disturb the natural habitat of rare species otherwise; we will lose most of the vulnerable orchids with potential horticultural importance.

REFERENCES

Alphonso, A. G. 1975. The role of the botanic gardens in the conservation of orchid species. Proceedings of the Eighth World Orchid Conference, 323–325. German Orchid Society Inc., Frankfurt.

Backhouse, G., Bramwells, H., Musker, R., Walker, G. and Lester, K. 1999. Frankston Spider orchid (*Caladenia robinsonii*, G.W. Carr). A Recovery Plan. 1999-2003.

Backhouse, G.N. and Jeans, J.A. 1995. The Orchids of Victoria. The Meigunyah Press, Melbourne.

Backhouse, G. and Cameron, D. 2005. Application of IUCN 2001 Red list categories in determining the conservation status of Native orchids of Victoria, Australia. *Selbyana.* 2(1): 58-74.

Bailes, C.P., 1985. Orchids in Nepal. *The Conservation and development of natural resources*. Advisory report and recommendations. Royal Botany Gardens, Kew.

Balakrishnan, N. P. 1981. Flora of Jowai. Vol. 1. Botanical Survey of India, Kolkata.

Bargali, K.U. and Joshi, M. 1998. Effect of forest covers on certain site and soil characteristics in Kumaun Himalayas. *Indian Journal of Forestry.* 21 (3): 224-227.

Borromeo, C. R. 1975. The urgent need for the conservation of Philippine orchid species. Proceedings of the Eighth World Orchid Conference, 341–342. German Orchid Society Inc., Frankfurt.

Briggs, J.D. and Leigh, J.H. 1995. Rare or Threatened Australian Plants. Revised Edition CSIRO Publishers, Melbourne.

Carr, G.W. 1991. New taxa in *Caladenia* R.Br. and *Gastrodia* R.Br. (Orchidaceae) from south eastern Australia. Miscellaneous Paper No. 1. Indigenous Flora and Fauna Association.

Das, A. K. and Khumbongmayum, A. D. 2006. *Galeola falconeri* Hook. f. an endangered giant saprophytic orchid. *Current Science.* 91(7): 871-873.

Gaur, R.D.1982. Dynamics of vegetation of Garhwal Himalayas. *The Vegetational Wealth of the Himalayas* (ed. G.S. Paliwal).12-25.

Ghanaksha, A. and Kaushik, P. 2007. Antibacterial Potential of Some Therapeutic Orchids. *J. Orchid Soc. India*, 21(1-2): 23-27.

Griesbach, R.J. 2002. Development of *Phalaenopsis* orchids for the mass-market. In: *Trends in new crops and new uses.* (ed. Janick, J. and Whipkey, A.) ASHA, Alexandria, USA.

Handa, S.S. 1986. Orchids for Drugs and Chemicals. In: *Biology Conservation and Culture of Orchids.* (ed. Vij, S.P.). Affiliated EW Press, New Delhi. pp. 89-100.

Husen, A. and Rahman, M.F. 2003. Orchids an Important Group of Plants for traditional systems of medicine in India. *Indian Forester.* 129(5): 651-653.

Kaushik, P. 1983. *Ecological and Anatomical Marvels of the Himalayan Orchids.* Today and Tomorrow Printers and Publishers, New Delhi. pp. 1-123.

Kumar M., Sharma C.M. and Rawat G.S. 2004. A study on the community structure and diversity of a sub-tropical forest of Garhwal Himalayas. *Indian Forester.* 130 (2): 207-214.

Nautiyal M.C., Nautiyal B.P. and Prakash V. 2004. Effect of grazing and climatic changes on alpine vegetation of Tungnath, Garhwal Himalaya, India. *The Environmentalist.* 24, 125–134.

Pattanaiki, C., Prasad, S.N. and Reddy, C.S. 2009. Need for conservation of biodiversity in Araku Valley, Andhra Pradesh. *Current Science.* 96(1): 11-12.

Pradhan, G. M. 1975. Habitat destruction of Himalayan orchid jungles. Proceedings of the Eighth World Orchid Conference, 331–334. German Orchid Society Inc., Frankfurt.

Ramakantha, V., Gupta, A.K. and Kumar, A. 2009. Biodiversity of Northeast India An overview. This is Google's cache of http://www.wii.gov.in/envis/rain_forest/chapter1.htm

Rogaly, J. M. 1975. Ecology and conservation of sub-tropical orchids of Southern Africa-mainly natal province. Proceedings of the Eighth World Orchid Conference. German Orchid Society Inc., Frankfurt.

Samant, S.S., Dhar, U. and Rawat R.S. 2001. In Himalayan Medicinal plants- potential and Prospects (eds Samant S, S, Dhar U and Palni, L.M.S.) Gyanodaya Prakashan. Nainital. 166-184.

Vij S.P., Srivastava, R.C. and Mainra A.K. 1992. On the occurrence of *Dactylorhiza hatagirea* (D.Don) Soo in Sikkim, Orhid News. 8-9: 14-15.

Xavier, T.F. and Senthilkumar, S. 2007. Management and conservation of medicinal orchids of Eastern Ghats of Tamil Nadu, South India. *J. Econ. Taxon. Bot.* 31(2): 417-422.

❑❑❑

Chapter 12

Phytoastrology : A Botanical Approach to Astrology

H.P. Pandey

INTRODUCTION

Plant worship has been an integral part of Indian civilization, culture and religion since antiquity. Our ancient scholars had explained the importance and relevance of this tradition in detail through contemporary literature like the *Vedas, Puranas, Samhitas* and *Smritis etc.* The plants are the only source of energy on this planet, which maintain all life forms and balance biosphere functioning, as they are capable of conversion of radiant energy of the sun into materials through photosynthesis. In this way, the sun is the lord of this planet and all life forms are maintained through plants. Therefore, all living beings on this earth are related directly or indirectly to the sun for their energy requirements.

It is estimated that the universe came into existence before some 15 billion years ago from the collision of matter and antimatter causing 'big bang' into *yleum,* a highly condensed substance of highest density. Due to big bang (greatest explosion) cloud of hydrogen-gas and related nuclei were produced, which after condensation and association gave rise to galaxies. According to astronomical perceptions, about 100 billion galaxies are moving apart at the speed of light (3 lakh km per second) giving expansion to the universe. Each galaxy consists of billions of stars in which millions of stars are dying and taking birth every moment. Our mother galaxy, the Milky Way is one of them, where the sun is an average type star constitutes its own family with planets and their satellites called as solar system, balanced with various stars, planets, galaxies of the universe through universal gravitational force. Thus, the whole universe is considered as an organic body, which has shared sensitivity or empathy.

Our mother planet earth originated from the sun and from which the entire living

---

Duthie Herbarium, Department of Botany, University of Allahabad, Allahabad, U.P.

world originated and evolved. It is therefore, concluded that the sun is maternal grandfather and plants and animals are brothers, as they have common origin form the mother earth. It has been scientifically proved that the substances (non-living and living) having common ancestry show internal isomerism and shared sensitivity or empathy. Since the body of animals and plants had originated from earth, which in turn is originated from the sun and sun had its origin from the universe, vis-à-vis plants, animals and microbes have mutual interdependence. Therefore, the life on earth is affecting and being affected by not only the solar system but also every object of the cosmos, based on the principles of empathy or shared sensitivity. However, the effect of solar system is more pronounced.

There are several instances in plant growth and movement, rhythmic and lunar cycles *etc.* An interesting example is that every eleventh year the annual rings of trees become more wide, and the radioactivity of the sun also increases enormously every eleventh year. It is common experience that the crime graph increases around full moon due to high gravitational pull exerted by moon on earth. It is therefore, obvious that if the plants in which nervous system is completely lacking are so much affected then the animals including human being, which have strong nervous and endocrine system for better coordination with environmental events are surely very much affected because of their unique capability of perceiving impulses and giving response against every event of the cosmos.

At cosmic level, all non-living and living objects have invisible relations, which can be analyzed and explained through metaphysics. This understanding and analysis of metaphysical relations of cosmos, solar system and their impacts on the life and well-being of humanity had been realized by ancient scholars, and conceived in the form of astrology. It has now become a vast science of diagnostic and remedial nature. According to astrology all the sufferings of mankind (individual or country) are due to planetary afflictions, which can be predicted earlier to suggest remedial measures to alleviate the sufferings if not completely neutralize it. The every aspect of this wide science of astrology is beyond the scope of this article. However, a brief preliminary knowledge about some important aspects like; Zodiac (*Rasis*), Constellations (*Nakshatras*), their lords and representative plants thereof are briefly discussed in proceeding paragraphs.

MATERIALS AND METHODS

The famous Indian astrologer *Varahmihir* designated only seven planets *viz.* the Sun, Moon, Mars, Mercury, Jupiter, Venus and Saturn. However, other scholars of astrology consider Rahu and Ketu also besides, seven planets. Thus, there are 9 planets, 12 zodiac and 27 constellations, which exert beneficial as well as malefic effects over the inhabitants of the earth including humans. Our great Indian scholars have laid foundation to balance the specific effects of these heavenly bodies and forces through *Mantra*, *Tantra* and *Yantra*. They have also suggested 9 specific gems as remedial measures to balance cosmic forces through colours and vibrations. Besides, various plants and trees, their parts and products, representatives of stars and planets have also been suggested to neutralize the malefic effects of respective cosmic bodies and forces.

Astrogardens

In ancient literature like *Narad puran, Narad samhita, Raj nighantu, Narayani samhita,*

Brihat sushrut, Sharda Tilak, Mantramaharnav and *Shrividyarnav* have mentioned astrological patterns of plantation, which not only subside the malefic effects of stars but also improve health, wealth and position of a person. This specific pattern can also help to remove negative effects of *vastu* and contribute a lot to improve the environmental quality and conservation of plant diversity. Based on gathered knowledge by the author himself, and literature survey following three types of astrogardens have been recorded.

1. Nine Planetary Garden (*Navgrah Vatika*)
2. Zodiac Garden (*Rasi Vatika*)
3. Constellary Garden (*Nakshatra Vatika*)

1. Nine Planetary Garden (*Navgrah Vatika*) – Indian astrologers have designated 9 planets which are the lords of various zodiac and constellations. A list of representative plants of every planet is given here.

(1) **The Sun:** *Artocarpus heterophyllus, Calotropis gigantea, Elaeocarpus sphaericus, Ficus racemosa, Ficus virens* and *Grewia subinaequalis.*

(2) **The Moon:** *Butea monosperma, Calotropis gigantea, Ocimum sanctum, Sapindus emarginatus* and *Syzygium cumini.*

(3) **The Mars:** *Acacia catechu, Aegle marmelos, Cocos nucifera* and *Prosopis cineraria.*

(4) **The Mercury:** *Achyranthes aspera, Cedrus deodara, Cynodon dactylon, Grewia tenax, Madhuca longifolia, Mesua ferrea* and *Pinus roxburghii.*

(5) **The Jupiter :** *Bambusa arundinacea, Desmostachya bipinnata, Ficus religiosa, Indigofera gerardiana, Mangifera indica* and *Mimusops elengi.*

(6) **The Venus:** *Butea monosperma, Ficus racemosa, Grewia subinaequalis Phyllanthus emblica* and *Saraca asoca.*

(7) **The Saturn:** *Azadirachta indica, Ficus religiosa, Mimusops elengi* and *Prosopis cineraria.*

(8) **The Rahu:** *Anthocephalus chinensis, Dalbergia sissoo, Terminalia arjuna* and *Terminalia bellirica.*

(9) **The Ketu :** *Bambusa arundinacea, Ficus benghalensis, Shorea robusta* and *Strychnos nux-vomica.*

Navgrah Vatika is very common astrogarden in almost every large temple even the astrologers establish this garden at their home with the help of pots, by growing representative plants of nine planets in following order either in circular or quadrangular piece of land (Fig. 1.0).

2. Zodiac Garden (*Rasi Vatika*) – The earth revolves around the sun in an imaginary circle of 360 degrees and if this circle is divided into twelve equal parts of 30 degrees, each segment with particular group of heavenly bodies is designated as zodiac signs or *Rasis*. Thus, we have twelve signs in the zodiac, namely – Aries (*Mesha*), Taurus (*Vrishabh*), Gemini (*Mithuna*), Cancer (*Karka*), Leo (*Simha*), Virgo (*Kanya*), Libra (*Tula*), Scorpio (*Vrischik*), Saggittarius (*Dhanu*), Capricorn (*Makar*), Aquarius (*Kumbha*) and Pisces (*Meena*). Table 1.0 provides a list of Zodiac signs, their lords and representative plants.

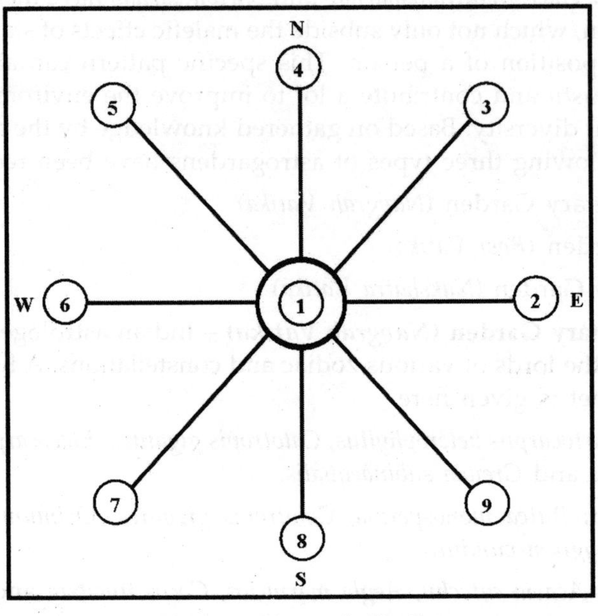

Fig. 1 : *Navgrah Vatika*

1. *Calotropis gigantea* (Aak), 2. *Ficus racemosa* (Gular), 3. *Achyranthes aspera* (Apamarg) 4. *Ficus religiosa* (Peepal) 5. *Desmostachya bipinnata* (Kush), 6. *Prosopis cineraria* (Shami) 7. *Cynodon dactylon* (Doob) 8. *Acacia catechu* (Khair), 9. *Butea monosperma* (Dhak).

In the garden the representative plant of zodiac sign (Table – 1.0) is planted in the east (1) and other plants are arranged in anticlockwise direction either in circular or quadrangular piece of land. A zodiac garden for Aries is given in Fig. 2.0.

TABLE 1

Zodiac signs	Lords	Representative plants	Vernacular names
1. Aries	Mars	*Acacia catechu*	Khair
2. Taurus	Venus	*Ficus racemosa*	Gular
3. Gemini	Mercury	*Achyranthes aspera*	Apamarg
4. Cancer	Moon	*Butea monosperma*	Palas
5. Leo	Sun	*Calotropis gigantea*	Aak
6. Virgo	Mercury	*Cynodon dactylon*	Doob
7. Libra	Venus	*Ficus racemosa*	Gular
8. Scorpio	Mars	*Acacia catechu*	Khair
9. Sagittarius	Jupiter	*Ficus religiosa*	Peepal
10. Capricorn	Saturn	*Prosopis cineraria*	Shami
11. Aquarius	Saturn	*Prosopis cineraria*	Shami
12. Pisces	Jupiter	*Desmostachya bipinnata*	Kush

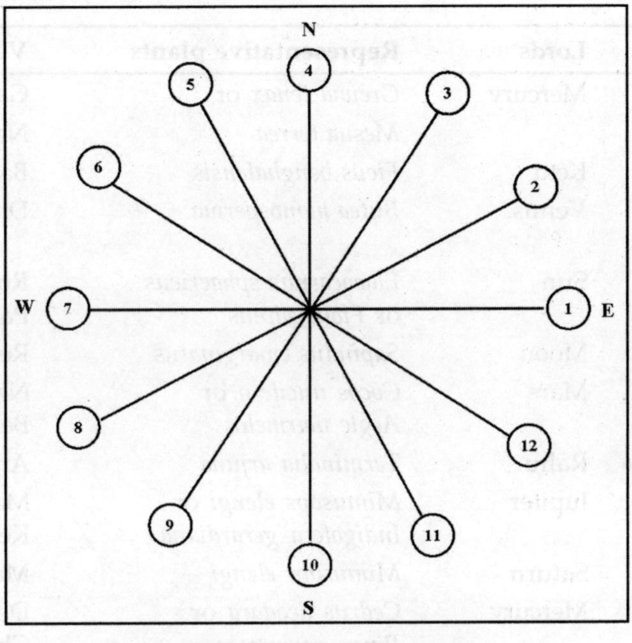

Fig. 2 : *Rasi Vatika*

3. Constellary Garden (*Nakshatra Vatika*) –The imaginary circle or path of earth is divided into 27 equal segments of 13°20' each to constitute a particular group of stars called as constellation or *Nakshatra*. Thus we have 27 constellations with the expansion of 13°20'. The following table 2.0 provides a list of constellations, their lords and representative plants.

TABLE 2

Nakshatras	Lords	Representative plants	Vernacular names
1. Aswini	Ketu	*Bambusa arundinacea* or *Strychnos nux-vomica*	Bans or Kuchla
2. Bharni	Venus	*Phyllanthus emblica* or *Grewia subinaequalis*	Aoı.wala or Phalsa
3. Kritika	Sun	*Ficus racemosa*	Gular
4. Rohini	Moon	*Ocimum sanctum* or *Syzygium cumini*	Tulsi or Jamum
5. Mrigshra	Mars	*Acacia catechu*	Khair
6. Aridra	Rahu	*Dalbergia sissoo* or *Terminalia bellirica*	Sisham or Bahera
7. Punarvasu	Jupiter	*Bambusa arundinacea*	Bans
8. Pushya	Saturn	*Ficus religiosa*	Peepal

Contd...

...Contd.

Nakshatras	Lords	Representative plants	Vernacular names
9. Aslesha	Mercury	*Grewia tenax* or *Mesua ferrea*	Gangerun or Nagkeshar
10. Magha	Ketu	*Ficus benghalensis*	Bargad
11. Purva falguni	Venus	*Butea monosperma*	Dhak, Palas
12. Uttara falguni	Sun	*Elaeocarpus sphaericus* or *Ficus virens*	Rudraksh or Pakar
13. Hasta	Moon	*Sapindus emarginatus*	Reetha
14. Chitra	Mars	*Cocos nucifera* or *Aegle marmelos*	Nariyal or Bel
15. Swati	Rahu	*Terminalia arjuna*	Arjun
16. Vishakha	Jupiter	*Mimusops elengi* or *Indigofera gerardiana*	Maulshri or Ketai
17. Anuradha	Saturn	*Mimusops elengi*	Maulshri
18. Jyestha	Mercury	*Cedrus deodara* or *Pinus roxburghii*	Deodar or Chir
19. Moola	Ketu	*Shorea robusta*	Shal
20. Purva ashada	Venus	*Saraca asoca*	Ashok
21. Uttara ashada	Sun	*Artocarpus heterophyllus* or *Grewia subinaequalis*	Kathal or Phalsa
22. Shravan	Moon	*Calotropis gigantea*	Aak
23. Dhanista	Mars	*Prosopis cineraria*	Shami
24. Satbhisha	Rahu	*Anthocephalus chinensis*	Kadamb
25. Purva bhadra	Jupiter	*Mangifera indica*	Aam
26. Uttara bhadra	Saturn	*Azadirachta indica*	Neem
27. Revti	Mercury	*Madhuca longifolia*	Mahua

This garden requires large piece of land to plant 27 plants and trees. The representative plants of each constellation (Table –2.0) either in circular or quadrangular piece of land starting from the east (1) *Aswini Nakshatra*, and other plants are arranged in anticlockwise direction, similar to that of *Rasi Vatika*. A single plant according to the *Nakshatra* in birth chart of descendent can also be planted anywhere, if the establishment of complete garden is not possible.

DISCUSSION AND RESULT

In Indian tradition of astrology, the solar system is considered as a super organic body (Kalapurush) and all the constituent planets as different parts that of, which govern the specific organs and systems of the kalapurush in twelve specific combinations, what is called as zodiac or rasis. *Varahmihir* designated seven planets *viz.* the Sun, Moon, Mars, Mercury, Jupiter, Venus and Saturn. The Sun constitutes soul of the kalapurish, the Moon his mind, Mercury the speech, Mars the body, Jupiter the knowledge, Venus the desire and Saturn his miseries. However, Rahu and Ketu are not substantial heavenly bodies with an observable mass and volume therefore called as shadow planets who affect in association with other planets.

A survey of astrological, ayurvedic, homoeopathic, medical, toxicogical and botanical literature (Ambasta, 1986; Anonymous, 1948-76; Anonymous, 1994; Basu. 1918; Bhambi, 2004; Bhandari, 1951-1957; Chopra, 1956; Chaudhuri & Pal, 1990; Dastur, 1952; Jain, 1967,1968, 1986,1991; Jain, & Mitra, 1990; Kapoor, 1998; Kirtikar. & Basu, 1918; Maheshwari, 1992, 1996; Pandey, 1999, 2000, 2001a, 2001b; Pandey & Verma, 2005; Powers, 1873-1875;Randhawa , 1983; Rastogi & Mehrotra, 1991; Sensarma, 1989; Wilson, 1961 etc.) reveal that the principal representative plants of planets (see Navgrah Vatika) coincide amusingly and surprisingly in their nature, characteristics and functions. However, astrology is a science of five dimensions therefore based on *panchang*, which needs to change our spectacle of science having the vision of only three dimensions. In future, the interdisciplinary studies related to astrology can reveal several prospects for the well being of humanity.

REFERENCES

Ambasta, S. P. (Ed.), 1986. *The Useful Plants of India.* Publication & Information Diretcrate, CSIR, New Delhi, India.

Anonymous, 1948-76. *The Wealth of India (Raw Materials).*Vol. 1-11, CSIR, New Delhi, India.

Anonymous, 1994. *Ethnobiology in India (A Status Report).* All India Coordinated Research Project on Ethnobiology (AICRPE), Department of Environment, New Delhi, India.

Basu, B. D. (Ed.), 1918. *Indian Medicinal Plants.* Indian Press, Allahabad.

Bhambi, A. 2004. *Be Your Own Astrologer.* Wisdom Tree, New Delhi.

Bhandari, C. 1951-1957. *Vanaushadhi Chandrodaya* (Hindi).10 Parts, Varanasi.

Chopra, R. N., S. L Nayar & I. C. Chopra, 1956. *Glossary of Indian Medicinal Plants.* Publication & Information Directorate, CSIR, New Delhi, India.

Chaudhuri, H. N. Rai & D. C. Pal, 1990. Plants in Folk-religion and Mythology. In S.K. Jain (Ed.) *Contribution to Ethnobotany of India* : 19-28, Scientific Publishers, Jodhpur.

Dastur, J. F. 1952. *Medicinal Plants of India and Pakistan.* Govt. Press.

Jain, S. K. 1967. Ethnobotany : Its scope and study. *Indian Museum Bull.* **2** (I): 39-43.

Jain, S. K. 1968. *Medicinal Plants.* National Book Trust, India.

Jain, S. K. 1986. Ethnobotany. *Interdisciplinary Science Reviews,* Vol. **II** (3): 285-292.

Jain, S. K. & R. Mitra, 1990. Ethnobotany In India : Retrospect and Prospect. In S.K. Jain (Ed.) *Contribution to Ethnobotany of India* : 1-17. Scientific Publishers, Jodhpur.

Jain, S. K. 1991. *Dictionary of Indian Folkmedicine and Ethnobotany.* Deep Publication, New Delhi.

Kapoor, G. S. 1998. *Remedial Measures in Astrology.* Ranjan Publications, New Delhi.

Kirtikar, K. R. & B. D. Basu, 1918. *Indian Medicinal Plants.* I-IV Vols., Bishen Singh Mahendra Pal Singh, Dehradun.

Maheshwari, J. K. 1992. (Ed.) *Ethnobotany in India*. Scientific Publishers, Jodhpur.

Maheshwari, J. K.1996. (Ed.) *Ethnobotany In South Asia*. Scientific Publishers, Jodhpur.

Pandey, H. P. 1999. *Ethnobotanical Studies in Uttar Pradesh*. D. Phil. Thesis, University of Allahabad, India.

Pandey, H. P. 2000. Phytofumigation : An Excellent Indian Healing System. *J. Phytological Res.* **13** (1): 77-80.

Pandey, H. P. 2001a. Seed oil of *Butea monosperma*, A Sexual Toner and Contraceptive. *Ethnobotany* **13**: 118-120.

Pandey, H. P. 2001b. Prakrati, Sanrachana avam Sanrakshan ki Vaidik Sankalpana ka Lok Jaivikiya Adhyayan. *Sachitra Ayurveda (J. Ayu. & Health)*, **54** (4): 263-266.

Pandey, H. P. & B. K. Verma, 2005. Phytoremedial Wreath : A Traditional Excellence of Healing. *The Indian Forester*, **131** (3): 437-441.

Powers, Stephen, 1873-1875. Aboriginal Botany. *California Academy of Science Proceeding*, **5**: 373-379.

Randhawa M. S. 1983. *Flowering Trees*. National Book Trust, New Delhi.

Rastogi, Ram, P. & B. N. Mehrotra, 1991. *Compendium of Indian Medicinal Plants*. Vol. 2 Central Drug Research Institute, Lucknow.

Sensarma, P. 1989. *Plants in the Indian Puranas*. Naya Prakosh Publication, Calcutta.

Wilson. M. H. 1961. *Vishnu Purana*. Punthi Pustak, Calcutta.

❑❑❑

Chapter 13

Concurrent Growing of Green Manure Crops in Rice

ANITHA S.

Rice is the most unique crop with a very wide adaptability. It is the staple food for more than half of the global population, of which 65 percent is in India. Rice production and productivity face a critical decline today. Rice cultivation has become less profitable in recent years due to increasing production cost and plataeuing yield levels. High labour cost, particularly for labour intensive operations like crop establishment and weeding, significantly contributes to increased production cost. In recent years there is an acute shortage of labour and high increase in labour wages. Scarcity of labour during critical stages of farm operations leads to low productivity. The productivity of paddy field is further affected due to the non application of organic manures in view of their high cost and limited availability. As a result there is a decline in the nutrient use efficiency. This situation necessitates the introduction of agronomically efficient technologies to increase the yield and lower the cost by optimum use of nutrients, and labour and thus make the rice farming remunerative.

Direct seeding of rice, either dry sown or wet sown, in low lands is a cost effective method of crop establishment followed by farmers in many rice growing areas including Kerala. Excessive weed growth is a major constraint in direct seeded rice, leading to high production cost and low productivity. System of Rice intensification (SRI) is also attracting attention by the rice farmers because of the multiple benefit it offers such as high yield and factor productivity with water saving. Under this system also excessive weed growth is one of the major constraints due to intermittent irrigation practice followed under SRI. The productivity is further adversely affected due to the non application of organic manures in view of their high cost and limited availability.

Associate Professor (Agronomy), Campus Development, Vellanikkara, Kerala Agricultural University, KAU P. O. 680 656, Thrissur, Kerala.

There is a greater possibility of intercropping green manure crops during the early stage of rice crop with less interference on rice growth (Mathew *et al.*, 1991). Many studies have indicated that concurrent growing of green manure crops and its subsequent incorporation in dry and wet seeded rice is very effective in supplying the required quantity of organic manures to rice with minimum investment and this practice has the additional benefit of weed suppression. Intercropping a green manure crop in rice and its incorporation later obviates the need for separate land, labour, inputs and time for a green manure crop.

CONCURRENT GROWING OF GREEN MANURE CROPS IN DRY SEEDED RICE

In Kerala, direct seeding of rice is done either by dry seeding in semi-dry system or by wet seeding in puddle condition. The dry sown (semi-dry) system of rice cultivation is a unique and extensively adopted rainfed rice ecosystem in Kerala, which constitute more than 60 per cent of the area under rice during *Kharif* season. In this system, the early growth of rice, upto 30-40 days is in a dry soil environment and thereafter comes under submergence with the onset of south west monsoon. The absence of stagnant water during the initial 4-6 weeks cause serious problems in dry sown low land rice with regard to application of organic manures and weed management affecting its productivity adversely.

The availability of enough green manure is a constraint and allocating an exclusive time slot and land for raising green manure is a limitation voiced so far. With the objective of evolving a suitable system for raising green manures, without sacrificing any of the economic crops in a rice based system, the strategy of growing green manure crops along with rice as a dual culture has been mooted.

In dry seeded system, the system of concurrent growing can be effectively capitalized by sowing green manure crops such as cowpea or horse gram along with rice crop. When rice is raised by dibbling, green manure seeds at the rate of 20kg ha^{-1} can be raised by dibbling in between two rows of rice in finely prepared soil. When broadcasting of rice is adopted, green manure seeds can be broadcasted after the broadcasting of rice and then both the seeds are to be covered by ploughing. Concurrently grown green manure crops in dry seeded rice, can be incorporated at 45 days after sowing by self decomposition when the fields get flooded with the onset of monsoon. If there is a failure of monsoon and consequent delay in the stagnation of water in the fields, then the green manure crop can be incorporated by spraying 2,4-D @1.0 kg ha^{-1}. In line sown crop, green manure crops can be incorporated using a cono weeder. Cono weeder is a hand operated mechanical weeder used for uprooting and burying weeds or green manure crops in between the standing rows of rice in wet lands. It can be pulled by a single person very easily, through the space in between the rows. On moving the cono weeder through the field, the weeds/green manures get cut and trampled into the soil. At the time of cono weeding, the soil should be well soaked and there should be a small film of standing water.

Field experiments conducted at Agricultural Research Station, Mannuthy during 2004 to 2006 revealed the benefits of concurrent growing of green manure crops with dry seeded rice.

TABLE 1

Effect of concurrently grown green manure crops, methods of incorporation and levels of nitrogen on weed count m^{-2} of dry sown rice

Treatments	Total weed count 30 DAS			Total weed count 50 DAS		
	1st year	2nd year	Pooled	1st year	2nd year	Pooled
Green manure crops						
Horsegram	203.00	156.60	179.80	146.60	121.40	134.00
Cowpea	130.00	97.11	113.60	88.16	84.88	86.52
CD (0.05)	20.99	14.19	17.85	12.76	7.78	6.07
Methods of incorporation						
Conoweeding	177.20	125.50	151.30	116.60	96.66	106.60
2,4-D spray	162.70	127.10	144.90	82.50	93.75	88.12
Self decomposition	159.70	127.90	143.80	153.10	119.10	136.10
CD (0.05)	NS	NS	NS	15.63	9.52	7.44
Levels of nitrogen						
100% N	166.60	127.10	146.90	116.30	101.10	108.70
75% N	166.40	126.60	146.50	118.40	105.30	111.90
CD (0.05)	NS	NS	NS	NS	NS	NS
Rice alone	269.00	302.30	285.70	367.00	326.30	346.70

Concurrent growing of green manure crops significantly reduced the weed population (75%) and weeds dry matter (58%) than sole cropping of rice (Table 1 and 2). Between the two green manure crops, cowpea was more effective in suppressing weeds and recorded the minimum weed population and weed dry matter. The reduction in weed growth evidently was due to the successful smothering effect of cowpea. Similar results were reported by Musthafa (1995) and Resmy (2003).The reduction in weed population and weed dry matter in the system of concurrent growing of green manure crops may be attributed to shading effect and competition stress created by the canopy of green manure crops in a unit area having suppressive effects on associated weeds, thus preventing the weeds to attain full growth. However the higher weed population and weed dry matter in horse gram intercropped plots compared with cowpea might be due to the slow growth of horse gram, providing more conducive conditions for growth of weeds than cowpea. The system of concurrent growing of green manures resulted in the reduction of weed population and thereby a labour saving of 40 man-days ha^{-1} for weeding alone. This reduced the cost of weed management substantially. The decrease in the population and dry matter production of weeds brought about by concurrent growing of green manure crops pointed out that green manure crops had grown only at the expense of weeds by using the growth resources which weeds would have otherwise utilized and may not have competed with rice for growth resources. Reduced weed population in green manure intercropped treatments appreciably reduced the competition for growth resources for rice and might have resulted in increased nutrient uptake and yield of rice compared with pure crop of rice.

TABLE 2

Effect of concurrently grown green manure crops, methods of incorporation and levels of nitrogen on dry weight of weeds (g m^{-2}) of dry sown rice

Treatments	Total weed dry weight 30 DAS			Total weed dry weight 50 DAS		
	1st year	2nd year	Pooled	1st year	2nd year	Pooled
Green manure crops						
Horsegram	16.30	15.85	16.08	22.84	17.53	20.19
Cowpea	12.92	12.34	12.63	16.13	15.00	15.56
CD (0.05)	0.73	0.59	0.67	1.12	0.51	2.20
Methods of incorporation						
Conoweeding	14.38	14.13	14.25	17.94	16.21	17.07
2,4-D spray	14.77	14.08	14.43	15.52	15.99	15.76
Self decomposition	14.68	14.08	14.38	25.00	16.59	20.80
CD (0.05)	NS	NS	NS	1.37	NS	2.56
Levels of nitrogen						
100% N	14.71	13.97	14.34	19.13	16.27	17.70
75% N	14.51	14.22	14.36	19.84	16.25	18.05
CD (0.05)	NS	NS	NS	NS	NS	NS
Rice alone	21.52	18.24	19.88	49.88	23.50	36.69

Concurrent growing of green manure crops and its subsequent incorporation in dry seeded rice is very effective in supplying the required quantity of organic manures to rice with minimum investment i.e., the cost of green manure seeds and its sowing and incorporation charges alone. It was observed that between the green manure crops tried, cowpea resulted in higher biomass addition compared to horse gram. Incorporation of cowpea resulted in a green matter addition of about 12 t ha^{-1} and was almost three times to that of horse gram (Table 3). It is a clear reflection of increased growth characters viz., plant height and leaf area by cowpea, over horse gram that has contributed to its quick growth rate in the early stages. These results are in agreement with that of Bridigit *et al.* (1994). The green matter addition by cowpea has been more than sufficient to meet the requirement 5 t ha^{-1} of organic manure recommended for rice in the state of Kerala (KAU, 2002). Another advantage of the system is that no separate land or time is required for raising the green manure crop. Biomass production of 12 t ha^{-1} utilizing the rainfall received only from the pre-monsoon rains also indicate that the system is viable and is capable of meeting the organic matter requirement of rice with minimum investment on water. The dry matter addition due to cowpea was 1.6 t ha^{-1} and almost double to that of horse gram. The corresponding nutrient addition by cowpea was 30 kg N ha^{-1}, 4 kg P ha^{-1} and 40 kg K ha^{-1}. These results confirm the finding of Beri *et al.* (1989), John *et al.* (1989) and Maskina *et al.* (1989). The contribution of N, P and K by cowpea was almost double than that of horse gram and there by saving 25 per cent of nitrogen fertilizer to the rice crop. Similar was the findings of Resmy (2003). Concurrent growing of cowpea significantly increased the nutrient uptake of rice over pure crop of rice. Nutrient content of plants at any time is a function of availability and is a factor for growth where as nutrient uptake is a function of growth, absorption and accumulation (Musthafa, 1995). The content of N, P and K was found to be higher in rice intercropped with green manure crops compared to pure crop of rice. The increased nutrient content of rice can be attributed to the improvement of soil fertility status and better nutrient availability due to the incorporation of green manure crops over control. Incorporation of green manure crops released nutrients slowly on their decomposition which resulted in better crop growth and higher dry matter production. Higher total N, P and K uptake at harvest due to green manure incorporation might be attributed to higher N, P and K content in plants together with higher dry matter production over the control treatment. Higher nutrient content of rice in the system of concurrent growing of green manure crops resulted in better performance of growth and yield attributes and there by higher yield of rice grown along with green manure crops over pure crop.

TABLE 3

Biomass production and dry matter addition of green manure crops grown along with dry sown rice at the time of incorporation

Treatments	Biomass (kg ha^{-1})			Dry matter (kg ha^{-1})		
	1st year	2nd year	Pooled	1st year	2nd year	Pooled
Horsegram	5181	2366	3774	849	821	835
Cowpea	15639	9066	12352	1883	1372	1628
CD (0.05)	1091	509	576	134	67	72

Concurrent growing of green manure crops along with dry seeded rice and its subsequent incorporation had no adverse effect on the growth and development of rice at any of the growth stages. Concurrent growing of green manure crops significantly increased the yield of rice compared to pure crop of rice. Increase in yield was brought about by the improvement in better growth parameters like height, tillers, leaves, leaf area, leaf area index and dry matter production and thereby yield contributing characters like number of panicles, filled grains and 1000 grain weight of rice, in the system of concurrent growing of green manures. Concurrent growing of green manure crops along with dry seeded rice and its subsequent incorporation has led to a yield increment of 11 per cent compared to pure crop of rice (Table 4). Similar results were obtained by Mathew *et al.* (1991) and Bridgit *et al.* (1994). Rice, in association with cowpea recorded a yield of 3633 kg ha^{-1}, which was six per cent more than the yield of rice realized in association with horse gram. It was observed that all the growth parameters of rice were superior in concurrent growing of cowpea. Rice in rice +cowpea system produced more height, tillers, leaves, leaf area and leaf area index than rice in rice+ horse gram system. Jayachandran and Veerabadran (1996) also reported an increase in plant height, leaf area index and total tillers of semi-dry rice as a result of incorporation of intercrops. Though not significant, the better tillering of rice coupled with higher leaf area helped in tapping more solar radiation and have resulted in better dry matter production in rice by concurrent growing of cowpea than the total dry matter production of rice in rice+ horse gram system and pure crop of rice. Better growth parameters had a positive reflection on yield contributing characters, like number of panicles, panicle weight, filled grains and 1000 grain weight of rice, in the system of concurrent growing of green manures compared with pure crop of rice. Musthafa (1995) reported a similar increase in grain and straw yield of rice due to the increase in productive attributes viz., productive tillers and 1000 grain weight, brought about by the interacting influences of fertilizer N and cowpea intercropping. It is observed in the present study that the better performances of growth and yield contributing characters resulted in higher yield of rice when intercropped with green manure crops. Pure crop of rice recorded lesser height, tillers and leaf area than green manure intercropped rice resulting in poor development of photosynthetic surface. This has lead to poor expression of yield contributing characters like number of panicles, panicle weight, filled grains and test weight. All these have resulted in 11 per cent reduction in yield of pure crop of rice compared with green manure intercropped rice.

TABLE 4

Effect of concurrently grown green manure crops, methods of incorporation and levels of nitrogen on grain yield and straw yield of dry sown rice

Treatments	Grain yield (kg ha^{-1})			Straw yield (kg ha^{-1})		
	1st year	2nd year	Pooled	1st year	2nd year	Pooled
Green manure crops						
Horsegram	3420	3426	3423	2230	3102	2666
Cowpea	3611	3656	3633	2293	3156	2725
CD (0.05)	NS	NS	210	NS	NS	NS
Methods of incorporation						
Conoweeding	3624	3547	3585	2294	3167	2730
2,4-D spray	3456	3570	3513	2229	3138	2683
Self decomposition	3467	3506	3486	2263	3082	2672
CD (0.05)	NS	NS	NS	NS	NS	NS
Levels of nitrogen						
100% N	3532	3609	3570	2261	3135	2698
75% N	3499	3473	3486	2263	3122	2693
CD (0.05)	NS	NS	NS	NS	NS	NS
Rice alone	3315	3220	3268	2361	3210	2785

The high yield obtained from concurrent growing of green manure crops along with rice suggested that in a farming situation, where application of green manure is not possible at the time of crop commencement, green manure crops can be grown during the early stage of rice crop with little interference on rice growth but with substantial yield advantage and without sacrificing any of the economic crops in the system. The viability of concurrent raising of green manure crops in closely seeded crops like rice largely depend on the possibility of competition it may offer to the main crop on growth resources. It is observed in the present study that, the better performances of growth and yield contributing characters were noticed when rice was grown along with green manure crops than when grown alone. This implied that competition to rice due to concurrent growing of green manure crops was almost absent. In addition green manure crops are leguminous plants that can fix atmospheric N for its need and as such may not compete with rice. The grain yield increase to the tune of 365 kg ha^{-1} obtained due to concurrent growing of green manure crops would imply that easily decomposable green manures can be applied even up to about 45 days of sowing of rice crop. Thus concurrent growing of green manure crops is especially advantageous to dry seeded rice crop as no green manures can be applied at the time of sowing because the soil is dry.

System of concurrent growing of green manure crops in dry seeded rice is economically profitable with an increase in net return by four times when rice is grown along with cowpea (Rs.19133 ha^{-1}) and two times with horse gram (Rs.11991 ha^{-1}) compared with pure crop of rice (Rs.4737 ha^{-1}). The net return and B: C ratio was not significantly affected by the methods of incorporation. Concurrent growing of green manure crops in dry seeded rice could save 25 per cent N to the current season rice crop with positive effects on the productivity and profitability of the system.

Concurrent growing of green manure crops during the previous season significantly increased the yield of succeeding transplanted rice crop compared to rice grown alone during the previous season. The soil fertility status after the succeeding crop indicated a positive balance of all the nutrients in treatments involving concurrent growing of green manure crops during the preceding season. The highest system profitability is obtained with concurrent growing of cowpea during the preceding season. Increase in net return of rice+ cowpea – rice system was Rs.9269 ha^{-1} compared with rice+ horse gram – rice and Rs.18856 ha^{-1} compared with rice- rice with out green manure crops.

Concurrent growing of cowpea along with dry seeded rice is a viable system as it resulted in increased productivity, profitability and sustainability dry seeded rice.

CONCURRENT GROWING OF DAINCHA IN WET SEEDED RICE

Wet sowing is a system of crop establishment in rice cultivation where pre-germinated seed is broadcast into puddle soil in areas with effective water control in the initial period. In wet seeded rice, the system involves raising of daincha (*Sesbania aculeata*) as a green manure crop concurrently with wet seeded rice. It can be done either by broadcasting daincha at the rate of 20 kg ha^{-1} after the broadcasting of rice at the rate of 60 kg ha^{-1}, or by sowing of rice (60 kg ha^{-1}) and daincha (20kg ha^{-1}) simultaneously into puddled soil using a rice-cum-green manure seeder. Sprouted rice seeds, with radicle just emerging, and un-sprouted daincha seeds are to be used for sowing. For sowing rice using rice cum green manure seeder the seeds are to be first soaked in water for 24 hours and then incubated for 12 hours. Then the seeds are filled in the 'Paddy seed hoppers' of rice-cum-green manure seeder to about three forth volume to allow free flow of seed through the holes. Un-sprouted green manure seeds are filled in the 'Green manure seed hoppers' of rice-cum-green manure seeder. When drawn in the field, the rice-cum-green manure seeder places sprouted paddy seeds and un-sprouted green manure seeds in alternate rows. Sowing is done when there is no standing water in the field. The daincha attains sufficient size by about 30 days after sowing and is then incorporated either by spraying 2, 4-D @1.0 kg ha^{-1}, chlorimuron ethyl + metsulfuron methyl @ 4.0 g ha^{-1} *or* metsulfuron methyl @ 5 .0 g ha^{-1}. In drum seeded crop, green manure crops can be incorporated using a cono weeder or by spraying weedicides.

Field experiments conducted at the Agricultural Research Station, Mannuthy during 2004 to 06 revealed that concurrent growing of daincha and its subsequent incorporation at 30 DAS was efficient in supplying the required green matter to rice with weed suppression and with out affecting the yield of wet seeded rice.

TABLE 5

Effect of stages and methods of incorporation of concurrently grown daincha and nitrogen levels on weed count m^{-2} and dry weight of weeds (g m^{-2}) of wet sown rice at 50 DAS

Treatments	Weed count m^{-2}			Dry weight of weeds (gm^{-2})		
	1st year	2nd year	Pooled	1st year	2nd year	Pooled
Stages of incorporation						
20 DAS	26.94	44.44	35.69	19.89	29.75	24.82
30 DAS	27.72	40.67	34.19	20.13	30.43	25.28
CD (0.05)	NS	NS	NS	NS	NS	NS
Methods of incorporation						
Conoweeding	36.25	50.00	43.13	20.29	31.53	25.91
2,4-D spray	19.08	36.00	27.54	18.69	28.94	23.82
Almix/Algrip spray	26.67	41.67	34.17	21.56	29.81	25.43
CD (0.05)	8.76	10.42	2.56	2.06	NS	NS
Levels of nitrogen						
100% N	27.06	41.44	34.25	19.03	29.18	24.10
75% N	27.61	43.67	35.64	21.00	31.01	26.00
CD (0.05)	NS	NS	NS	1.68	NS	NS
Rice alone	123.67	127.67	125.67	50.63	64.97	57.80

Concurrent growing of daincha along with wet seeded rice significantly reduced the count and dry weight of grasses, broad leaved weeds and sedges and thereby the total population and dry matter production of weeds compared to wet sown rice

(Table 5). The reduction in weed population and weed dry matter in the system of concurrent growing of daincha may be attributed to shading effect exerted by the canopy of daincha. The co-cropping system reduced the weeding cost by 59 per cent due to saving in labour requirement by 38 man-days ha^{-1}. Incorporation of daincha at 20 and 30 DAS had no significant influence on the weed population and weed dry matter production. The decrease in the population and dry matter production of weeds brought about by concurrent growing of daincha indicated that daincha had grown only at the expense of weeds by using the growth resources which weeds would have otherwise utilized and may not have competed with rice for growth resources. Reduced weed population in treatments involving co-cropping of daincha created a competition free environment for growth resources for rice and might have resulted in increased nutrient uptake and yield of rice compared with pure crop of rice.

TABLE 6
Biomass production and dry matter addition of daincha grown along with wet sown rice

Treatments	Biomass ($kg\ ha^{-1}$)			Dry matter ($kg\ ha^{-1}$)		
	1st year	2nd year	Pooled	1st year	2nd year	Pooled
Stages of incorporation						
20 DAS	5424	5864	5644	2224	3079	2651
30 DAS	9754	17643	13698	5501	6637	6069
CD (0.05)	938	1537	856	280.4	243	176

Concurrent growing of daincha and its subsequent incorporation in wet seeded rice is very effective in supplying the required quantity of organic manures to rice with minimum investment i.e. the cost of daincha seeds and its sowing charges alone. The green biomass addition at 30 DAS was 2.24 times more compared to incorporation of daincha at 20 DAS (Table 6). As suggested by Bhuiya and Salam (2002), the yield increase in rice was due to the incorporation of larger quantity of green biomass of *Sesbania*. Rajendran *et al.* (2002) opined that concurrent growing of daincha and rice using rice cum green manure seeder and the subsequent incorporation of daincha using conoweeder, yielded seven to nine tones of biomass per hectare within four to five weeks. The higher green biomass addition resulted in increased availability of nutrients especially nitrogen. This might be due to the increased nutrient content of daincha at 30 DAS than at 20 DAS. Higher biomass of daincha (13698 kg ha^{-1}) coupled with increased nutrient content significantly increased the nutrient contribution from daincha at 30 days of incorporation compared with incorporation at 20 days. Higher biomass production of daincha at 30 DAS was a positive reflection of increased height, leaf area and nodule count. The N, P and K contribution of 30 days old daincha was to the tune of 126.4, 8.86 and 139.2 kg ha^{-1} and the corresponding figures of 20 days old daincha was 40.1, 5.57 and 63.5 kg ha^{-1}. Singh *et al.* (1991a) reported that biomass production and N accumulation of *Sesbania* were mainly controlled by age factor. The increased growth of daincha led to better soil fertility and thus gave boost to yield attributes which in turn resulted in increased grain yield (Sah *et al.*, 2000). Increase in yield due to application of daincha might be due to addition of extra nutrients. Application of daincha exploits the fixed nutrients of soil in available form and regulates its supply to the crop through mineralization and prevents them from leaching and other losses (Dwivedi and Thakur, 2000).

N, P and K content and uptake of rice with daincha was higher than rice grown alone. Incorporation of daincha at 30 DAS resulted in build up of soil organic carbon and registered the highest available N, P and K in the soil compared with incorporation of daincha at 20 DAS and pure crop of rice. Increased organic matter content and N, P and K content of post harvest soil is due to the increased biomass addition and thereby increased nutrient contribution resulted by the incorporation of daincha.

TABLE 7

Effect of stages and methods of incorporation of concurrently grown daincha and nitrogen levels on grain yield and straw yield of wet sown rice

Treatments	Grain yield (kg ha^{-1})			Straw yield (kg ha^{-1})		
	1st year	2nd year	Pooled	1st year	2nd year	Pooled
Stages of incorporation						
20 DAS	5121	4694	4907	2887	4457	3672
30 DAS	5417	4951	5184	3138	4313	3725
CD (0.05)	203.3	239.1	202.3	239.2	NS	NS
Methods of incorporation						
Conoweeding	5319	4912	5115	2873	4453	3663
2,4-D spray	5265	4801	5033	3014	4456	3735
Almix/Algrip spray	5222	4754	4988	3150	4245	3697
CD (0.05)	NS	NS	NS	NS	NS	NS
Levels of nitrogen						
100% N	5384	4947	5165	3076	4317	3697
75% N	5153	4698	4925	2949	4452	3700
CD (0.05)	203.3	239.1	202.3	NS	NS	NS
Rice alone	4895	4107	4501	2932	4368	3650

Beneficial effect of concurrent growing of daincha in wet seeded rice was evident from the increased yield of rice compared to wet sowing of rice alone (Table 7). Stages of incorporation of daincha had significant influence on yield. Incorporation of daincha at 30 days after sowing resulted in significantly higher yield (5184 kg ha^{-1}) compared to 20 days after sowing (4907 kg ha^{-1}). It supports the findings of Jayadeva *et al.* (2004) who reported that inter cropping of daincha in drum seeded rice for *in situ* incorporation at 30 DAS and application of 100 per cent recommended N recorded significantly higher yield due to effective biomass incorporation into the soil. Incorporation of daincha at 30 DAS, though not significant, registered higher values of all the growth parameters compared with 20 DAS. These growth parameters positively reflected in the better expression of yield attributes and there by higher yield. It is observed in the present study that, the better performances of growth and yield contributing characters were noticed when rice was grown along with daincha than grown alone. Concurrent growing of daincha effectively minimized the production of non productive tillers compared with pure crop and this decline of non productive tillers in the post flowering phase had contributed to the increased yield (Table 4.3.7). This result is in close agreement with Musthafa (1995). This implied that competition for rice due to concurrent growing of daincha was less. In addition, daincha is a leguminous plant that can fix atmospheric N for its need and as such may not compete with rice. Concurrent growing of daincha in wet seeded rice proved better in terms of productivity over sole crop. The marked increase in grain yield could be owing to improvement in yield attributes and enrichment of soil fertility through green matter addition into the soil (Hemalatha *et al.*, 2000).

System of concurrent growing of daincha in wet seeded rice is economically profitable. Net return was found to be the highest when daincha was incorporated at the age of 30 days (Rs.27473 ha^{-1}) compared with daincha incorporation at 20 DAS. Concurrent growing of daincha and wet seeded rice and the subsequent incorporation of daincha is a low cost technology because it avoids the need for exclusive cultivation of green manure crop as well as the need for raising the nursery and transplanting, thus saving considerable labour and cost

Incorporation of daincha at 30 DAS during the preceding season registered significantly higher grain yield of succeeding rice than incorporation of daincha at 20 DAS. The soil fertility status after the succeeding crop indicated a build up of all the nutrients by the concurrent growing of daincha along with the preceding wet sown rice.

The highest system profitability is obtained by incorporating concurrently grown daincha at 30DAS. Increase in net return of the rice + daincha- rice system due to the incorporation of 30 days old daincha was Rs.5396 ha^{-1} compared to the incorporation of 20 days old daincha and Rs.17088 ha^{-1} compared with rice –rice system with out daincha.

Concurrent growing of daincha can be successfully practiced in wet seeded rice with improved productivity. Intercropping of daincha in wet seeded rice increased the profitability of the rice- rice cropping system.

CONCURRENT GROWING OF DAINCHA IN RICE UNDER SRI MANAGEMENT

Concurrent growing of daincha along with rice under SRI management was carried out to formulate strategies to control weeds and make available organic manure at cheaper cost.

Field experiment was conducted at Agricultural Research Station Mannuthy during 2008-09. Treatments comprising planting of seedling viz., 10 days old one seedling per hill at two spacing, viz., 25 x 25 cm and 20 x 15 cm; Irrigation viz., intermittent irrigation and continuous flooding; Weed management viz., cono weeding, manual weeding and by concurrent growing of daincha and its incorporation using 2, 4- D 1.0 kg ha^{-1} at 30 days after sowing. In cono weeded plots weeding starts at 10 days after planting and then 2 weeding at 10 days interval. In concurrently grown daincha plots, daincha seeds (20 kg /ha) were sown on the next day of planting 10 days old one seedling per hill and incorporated the daincha by spraying 2, 4- D 1.0 kg ha^{-1} at 30days after sowing.

Concurrent growing of daincha and subsequent incorporation of daincha by desiccating it at 30 days after sowing by spraying 2, 4 – D @ 1kg ha^{-1} was found to be very effective in controlling the weeds and supplying the required quantity of organic manures (12 t ha^{-1}) to rice with minimum investment without any yield reduction. Here, the investment is required only for the cost of daincha seeds and for the cost incurred for incorporating daincha. Incorporation of daincha at 30 DAS added 12 t ha^{-1} of green matter and it was also higher than the recommended dose of 5 t ha^{-1} of organic manure.

TABLE 8

Effect of concurrent growing of daincha in SRI on the weed count and weed dry matter production at panicle initiation stage

Treatments	Weed count				Weed dry matter (g)			
	Grasses	Broad leaved	Sedges	Total	Grasses	Broad leaved	Sedges	Total
Spacing								
25 x 25 cm	3.44	5.3	4.38	13.16	0.47	0.99	0.36	1.84
20 x 15 cm	3.22	7.3	4.44	15.00	0.44	1.05	0.40	1.90
CD at 0.05	NS	NS	NS	NS	NS	NS	NS	NS
Irrigation								
Intermittent irrigation	3.27	9.38	4.77	17.44	0.53	0.98	0.42	1.95
Continuous flooding	3.38	3.27	4.05	10.72	0.38	1.06	0.34	1.79
CD at 0.05	NS	3.83	NS	5.83	NS	NS	NS	NS
Weeding								
Cono weeding	4.66	8.33	5.83	18.83	0.69	1.21	0.48	2.39
Manual weeding	3.58	8.84	5.58	17.58	0.58	1.26	0.48	2.33
Concurrent growing of daincha	1.75	2.25	1.83	5.83	0.11	0.60	0.18	0.89
CD at 0.05	NS	4.69	2.42	7.11	0.35	0.54	0.27	0.70

The reduction in weed population and weed dry matter in the system of concurrent growing of daincha may be attributed to shading effect exerted by the canopy of daincha. No significant difference noticed in the grass count but count of broad leaved and sedges count was less in daincha grown plots. Concurrent growing of daincha and its incorporation using 2,4-D 1.0 kg ha^{-1} at 30DAS significantly reduced the population and dry matter production of broad leaved weeds and sedges. Incorporation of daincha by spraying 2,4-D resulted in 74 per cent reduction of total weed count and 64 per cent reduction in weed dry matter production (Table 8). This was mainly because of the combined effect of concurrent growing of daincha and its subsequent incorporation by spraying 2, 4 – D. Application of 2, 4-D for incorporation of daincha by desiccation controlled broad leaved weeds and sedges substantially because 2, 4-D is a selective herbicide recommended for broad leaved weeds and sedges in rice. As a result these treatments recorded the lowest count and dry matter of broad leaved weeds and sedges. Accordingly the highest reduction of total weed count and dry matter production was noticed when daincha was incorporated by 2,4- D spray compared to cono weeding and manual weeding plots.

Spacing had no significant influence on the weed count and weed dry matter production. Irrigation treatment had significant influence on weed count. Grasses and sedges count showed no significant difference in irrigation treatments. Broad leaved weeds and total weed count was significantly higher in intermittent irrigation treatment. Weed dry matter production was not influenced by irrigation treatment.

Concurrent growing of daincha had no significant influence on the growth parameters of rice under SRI. Plant height, tiller production, leaf area development and LAI and dry matter production was not affected by growing daincha along with rice. This indicated that concurrently grown daincha had no competition with rice for growth resources. The decrease in the population and dry matter production of weeds brought about by concurrent growing of daincha indicated that daincha had grown only at the expense of weeds by using the growth resources which weeds would have otherwise utilized and may not have competed with rice for growth resources. Concurrent growing of daincha did not significantly influence the yield parameters and yield of rice under SRI. Straw yield was significantly less in daincha grown plots. Harvest index was highest when daincha was grown along with rice.

Cost of cultivation was the least in daincha grown plots compared to other weeding treatments. By growing daincha the labour requirement for weeding alone was reduced by 44 man days under SRI management. Thereby there was a saving of labour cost for Rs. 3750 ha^{-1} for weeding alone. Net return and B: C ratio was significantly increased when daincha was grown along with rice.

Concurrent growing of green manure crops and its subsequent incorporation of by desiccating it at 30 days after sowing by spraying 2, 4 – D @ 1kg ha^{-1} is a cost effective weed management alternative for rice under direct seeded and SRI management. It is very effective in supplying the required quantity of organic manures to ric. with minimum investment.

It is concluded that resource poor farmers can get higher returns by concurrent growing of leguminous crops in rice based cropping systems, besides the sustainability in production and soil health.

REFERENCES

Beri, V., Meelu, O. P. and Khind, C. S. 1989. Biomass production, N accumulation, symbiotic effectiveness and mineralization of green manures in relation to yield of wet land rice. *Tropical Agriculture* **66**:11-16.

Bhuiya, M. S. U and Salam, M. A. 2002. Organic matter management through sesbania manuring in rice-rice cropping system. *Proceedings of the 17th WCSS*, 14-21 August 2002, Thailand. Paper No. 2017.

Bridgit, T. K., Potty, N. N. and Marykutty, K. C. 1994. A management model for the improvement of soil and crop productivity in semi-dry rice. *Proceedings of the Sixth Kerlala Science Congress*, 29-31 January 1994, Thiruvananthapuram. pp. 125-126.

Dwivedi, D. K. and Thakur, S.S. 2000. Effect of organics and inorganic fertility levels on productivity of rice (*Oryza sativa*) crop. *Indian Journal of Agronomy* **45(3)**: 568-574.

Hemalatha, M., Thirumurugan, V. and Balasubramanian, R. 2000. Effect of organic sources of nitrogen on productivity, quality of rice (*Oryza sativa*) and soil fertility in single crop wet lands. *Indian Journal of Agronomy* **45(3)**: 564-567.

Jayachandran, M. and Veerabadran, V. 1996. Nitrogen economy through intercropping green manures in semi-dry rice. *Madras Agriculture Journal* **83**:466-468.

Jayadeva, H. M., Denesh, G. R., Bhairappanavar, S. T. and Gowda, T. H. 2004. Role of daincha (*Sesbania aculeate*) for sustainable rice production. *Extended Summaries Vol.2: International symposium on Rice- From Green revolution to Gene revolution* (eds Bentur, J. S., MIlyas Ahamed., Rao, K. J. and Mishra, B.). 04-06 October 2004, Indian Council of Agricultural Research, New Delhi and Directorate of Rice Research, Rajendra Nagar, Hyderabad pp 319-320.

John, P. S., Pandey, R. K., Buresh, R. J. and Prasad, R. 1989. Lowland rice response to urea following three cowpea cropping systems. *Agronomy Journal.* **81**(6): 853-57.

KAU, 2002. *Package of Practices Recommendation Crops.* 12th edn. Kerala Agricultural University, Thrissur. 278 p.

Maskina, M. S., Singh, B., Singh, Y. and Meelu, O. P. 1989. Integrated nitrogen management with green manures in rice-wheat system. *Oryza* **26**:358-362.

Mathew, G., Alexander, D. and Bridgit,T. K. 1991. Intercropped cowpea is an ideal green manure for semi-dry rice. *Indian Farming* 41(2): 21-22.

Musthafa, K. 1995. Productivity of semi-dry rice under simulataneous in situ green manuring. M. Sc. (Ag.) thesis, Kerala Agricultural University, Thrissur. 143p.

Rajendran, P., Tajuddin, A. and Nadanassababady, T. 2002. Seeder developed for direct sowing of rice (*Oryza sativa*) and green manure under puddle soil surface. *Extended Summaries Vol.1 2nd International Agronomy Congress*, 26-30 November 2002, Indian Council of Agricultural Research, New Delhi. pp. 383-384.

Resmy, O. N. 2003. Weed management in semi-dry rice intercropped with green manure crops. M. Sc. (Ag.) thesis, Kerala Agricultural University, Thrissur. 83 p.

Sah, R. R., Rai, R. K. and Mukherjee, P. K. 2000. Effect of green-manure dhaincha (*Sesbania aculeate*) and phosphorus on growth, yield and phosphorus uptake by wheat (*Triticum aestivum*) *Indian Journal of Agronomy* **45(4):** 707-710.

Singh, D., Pal, D. and Om, H. 1991. Effect of age and period of decomposition of daincha on the yield of dwarf rice. *Haryana Journal of Agronomy* **7**: 89-90.

❑❑❑

Chapter 14

Indian Forests and Research—Past and Present

S.K. BANERJEE* AND AVINASH JAIN**

INTRODUCTION

India is the seventh largest country in the world and Asia's second largest nation with an area of 3,287,263 square km. The Indian main land stretches from 6°45′ to 37°6′ N latitude and from 68°7′ to 97°25′ E longitude. It has a land frontier of some 15,000 km and a coast line of 7,516 km. There exists complete rainless regions to world's highest rainfall area. The climate ranges from tropical to sub tropical and temperate to arctic.

India possesses a distinct identity, not only because of its geography, history and culture but also because of the great diversity of its natural ecosystems. The panorama of Indian forests ranges from evergreen tropical rain forests in the Andaman and Nicobar islands, the Western Ghats, and the north-eastern states, to dry alpine scrub high in the Himalayas to the north. Between the two extremes, the country has semi-evergreen rain forests, deciduous monsoon forests, thorn forests, sub-tropical pine forests in the lower montane zone and temperate montane forests (Lal 1989).

One of the most tropical forests classifications was developed for Greater India (Champion 1936) and later republished for present day India (Champion and Seth 1968). In it 16 major forest types are recognized (Table 1), and subdivided into 221 minor types. Structure, physiognomy and floristic are all used as characters to define the types.

There is no doubt that forests contain more biodiversity, so far as macro-flora and fauna are concerned in comparison to any other ecosystem. India is a mega diverse country and houses not less than 45000 plant species and 75000 animal species. At least 7000

*Senior Scientist (Retd.), D-19 Rajul Paradise, P,O, Bilhari, Jabalpur 482,020 (MP)

**Ecology & Rehabilitation Division, Tropical Forest Research Institute, P.O. RFRC, Mandla Road, Jabalpur 482 021 (MP)

TABLE 1
Major forest types in India

Sr. No.	Forest type	Area (million ha)	% of forest area	Occurrence in States/UTs
1.	Tropical wet evergreen forest	4.5	5.8	Arunachal Pradesh, Assam, Karnatak Kerala, Manipur, Nagaland, Tamil Nadu, Andaman & Nicobar Islands and Goa
2.	Tropical semi-evergreen forest	1.9	2.5	Assam, Karnataka, Kerala, Maharashtra, Nagaland, Tamil Nadu Andaman & Nicobar Islands & Goa
3.	Tropical moist deciduous forest	23.3	30.3	Andhra Pradesh, Assam, Bihar, Gujarat,Karnataka, Kerala, M.P., Maharashtra, Manipur, Meghalaya, Mizoram, Tripura, Nagaland, Orissa, Tamil Nadu
4.	Littoral and swamp forest	0.7	0.9	Andhra Pradesh, Gujarat, Orissa, Maharashtra, Tamil Nadu, W.B., Andaman 7 Nicobar Islands
5.	Tropical dry deciduous forest	29.4	38.2	A.P., Bihar, Gujarat, Haryana, Himachal Pradesh, Kerala, Karnatak M.P., Maharashtra, Jammu & Kashm Punjab, Rajasthan, Tamil Nadu, U.P.
6.	Tropical thorn forest	5.2	6.7	A.P., Gujarat, Haryana, H.P., Karnata M.P., Maharashtra, Punjab, Rajasthar Tamil Nadu, U.P.
7.	Tropical dry evergreen	0.1	0.1	A.P., Tamil Nadu
8.	Sub-tropical broad leaved hill forest	0.3	0.4	Assam, Meghalaya
9.	Sub-tropical pine forest	3.7	5.0	Arunachal Pradesh, Manipur, Sikkim Meghalaya, Nagaland, H.P., Jammu& Kashmir, Haryana, U.P., Punjab
10.	Sub-tropical dry evergreen forest	0.2	0.2	H.P., Jammu & Kashmir
11.	Montane wet temperate forest	1.6	2.0	Arunachal Pradesh, Manipur, Sikkim Nagaland, Karnataka, Tamil Nadu
12.	Himalayan moist temperate forest	2.6	3.4	H.P., Jammu & Kashmir, U.P.
13.	Himalayan dry temperate forest	0.2	0.2	H.P., Jammu & Kashmir
14.	Sub-alpine forest	-	-	Arunachal Pradesh, H.P.
15.	Moist alpine-scrub	3.3	4.3	Jammu & Kahmir, U.P.
16.	Alpine scrub	-	-	Jammu & Kashmir, U.P.

of these plant species are indigenous. The flora of India is greatly influenced by Indo-Malaysian, Eurasian and Tropical –African vegetation.

A geographical zone or ecological niche with a large number of endemic plant is called a Hot spot. Mcneely *et al.* (1990) estimated that 70% of world's total flowering plants occur in 12 countries and these have been designated as Mega diversity centres or Mega diversity countries. India gets the 10th place in the world and 4th in Asia.

In India there are three mega endemic centres.

(1) Eastern Himalayas : 1808 endemic plants from approximately 6000 species.

(2) Western Ghats : 1500 endemic plants from 4000 species.

(3) Western Himalayas : 1195 endemic plants from 5000 species.

There are eight subcentres identified within our country, namely Western Himalaya, Eastern Himalaya, North Eastern Regions, Gangetic Plains, Indus Plains, Western Ghats, Eastern Ghats and Andaman and Nicobar Islands.

The floristic richness has also rendered the North-Eastern Region and Western Ghats to be recognized as two hot spots among 18 hot spots identified throughout the world. In addition to these two, there are about 40 other sites in different phyto-geographical zones of India which have high degree of endemism and genetic diversity (Table 2).

TABLE 2

Phyto-geographical zones of high degree of genetic diversity

1.	Andaman group of islands
2.	Nicobar group of islands
3.	Agasthyamalai hills
4.	Anamalai and High Ranges (Cardamom hills)
5.	Palni and High Way Mountain
6.	Niligiris, Salient Valleys-Wyanad-kodagu
7.	Shimoga-Kanara
8.	Mahabaleswar-Khandala Ranges
9.	Konkan-Raigad
10.	Marathwada Satpura Ranges
11.	Tripura-Cuddappa-Nallamalai Hills
12.	Vizagapatnam-Ganjam-Jeypore Hills
13.	Southern Deccan (Leeward side)
14.	Chtanagpur plateau
15.	Kathiawar Kutch

Contd...

...Contd.

16. Rahasthan-Aravalli
17. Khasia-Jaintia Hills
18. Patkoi-Manipur-Lushai Hills
19. Assam
20. Arunachal Pradesh Himalaya
21. Sikkim Himalaya
22. Garhwal Kumaon Himalaya
23. Lahul-Himachal {radish Himalaya
24. Kashmir Ladak Himalaya
25. Nepal (interface of E. Himalaya & W. Himalaya)
26. E. Himalaya Plant gene pool
27. Khasia-Jaintia-Lushai Plant grne pool
28. Central Indian Plant gene pool
29. E. Ghats Plant gene pool
30. Southern E.W. Ghats plant gene pool
31. Northern W. Ghats Plant gene pool
32. W. Himalaya Plant gene pool
33. Sandstone flora of Dun and Mussorie
34. Myristica Swamps of Kerala
35. Sea grasses of Coromandal Coast
36. Mangroves of Sunderbans
37. Mangroves and coral reefs of Andamans
38. Wetland flora of Chilka Lake
39. Cold desert flora of Ladakh
40. Lakshadeep coral reefs and algal flora

HISTORICAL

The forestry research in India has by now a long history and described below: Aryans migrated to India between 2000 – 1000 B.C. from the west. At that time, people in this country were using stones, bronze and copper for different purposes. The Aryans introduced the use of iron for manufacture of agricultural implements and also carried on forest felling for their settlement. Vedas written about 1500 B.C. have described the forests. It is believed that the Rishis had knowledge of poisonous and medicinal values of herbs, shrubs and trees. They advocated against cutting of trees as they believed that this would result in poor rainfall. During Maurya period, information about forests was recorded in Kautilya's Arthashastra (321 B.C.), Indika of Megasthenes (305 B.C.), Mudrarakshas of Vishakadatta, inscriptions of Ashoka (273 B.C. – 236 B.C.) etc.In Chandra

Gupta Maurya's administration, a regular forest department existed, wherein Kupyadhyakshas (Superintendent of Forest Products) were appointed to be the Head of the Department who were assisted by Vanpalas or Forest Guards. It was their duty to increase the productivity of forests and to earn revenue, in addition to classifying trees and other constituents of vegetation etc. Forests were classified as reserved forests, forests donated to eminent Brahmins, and forests for the public. During Gupta period, forests were treated as main source of revenue.

The Muslim rulers, though fond of road side trees, gardens and shooting, were comparatively less interested in forest management. However, during the period of Emperor Akbar, different kinds of wood were weighed, and higher and heavier woods were recognized as indicated in Ain-I-Akbari.

The early British administration was under the impression that the forest wealth of India was inexhaustible. As supplies of first class oak timber from other countries became short in England, large quantities of teak from India were used for the British Admiralty's fleet. In 1800, a commission was appointed to enquire into the availability of teak in Malabar forest and felling of teak below 52cm girth was prohibited. In 1805, a Forest Committee was constituted, as a consequence of whose report the East India Company declared "Royalty" rights over teak trees in South India. In 1806, Madras Government appointed the First Conservator of Forests to organize the production of teak and other timber suitable for ship building. In 1842, action was started for raising teak plantations in Malabar. In 1850, the British Association in Edinburgh formed a committee to consider the destruction of tropical forests in India which reported that over a large portion of Indian Empire, there was uncontrolled destruction of forests, whereas, in Malabar, improvements have taken place which needed to be extended to the forests over the rest of the country. The committee also stated that the forest felling around localities from which supply was to be derived was to be greatly deprecated.

In 1855, Lord Dalhousie, the then Governor General promulgated for the first time an outline for the forest conservancy for the whole country. In 1856, Dr. Brandis, a trained forester from Germany, was appointed Superintendent of Forests in Peru (Myanmar) who later came to serve India as Inspector General of Forests. Indian Forest Act of 1865 came into being as the first attempt at forest legislation, under which the local Governments were empowered to draft local rules for enforcement in their regions. Steps, accordingly, were taken to prevent acts which caused injury to the forests. Dr. Brandis took steps to recruit personnel for the establishment of Forest Department. Training of forest service probationers was arranged by Brandis in Europe. In 1867, five candidates were sent to France and two to Hanover in Germany. Later, responsibilities of training of IFS officers was given to the Universities of Oxford, Cambridge and Edinburgh. In 1870, the foundation of forest department had been truly laid and definite progress in forestry was obvious. Systematic forestry research in the country made an early beginning sometimes in 1878 when research on utilization of forest products commenced from incipient chemical laboratory established in the forest school at Dehra Dun. In 1878, the revised Forest Act was passed. In the same year, a Forest School was inaugurated at Dehra Dun for training Forest Rangers. During the year 1891, a Provincial Forest Service was established to recruit and train in India itself suitable persons to work as Forest Officers. From 1926 to 1932

IFS Probationers were trained at Dehra Dun after which the service was discontinued. The recruitment to IFS was again started in 1966, and the training of regular recruits of the service was resumed from 1968 at the Indian Forest College, which was converted into Academy in the year 1987.

In his report on "Improvement of Indian Agriculture", 1893, Dr. Voelcker laid stress on the need for formulating a Forest Policy with a definite bias for serving the agricultural interest. Accordingly in 1894, a Forest Policy was formulated which brought out that forests be managed to promote the general well being of the people and be maintained for the preservation of climate and physical conditions of the country. In this policy, permanent cultivation was put before forestry and satisfaction of the need of local population was given over-riding consideration over state revenue. The Government owned forests were classified into Protection Forests, Commercial Forests, Minor Forests and Pasture Lands.

In 1906, the Imperial Forest Research Institute (IFRI) was established at Dehra Dun with six scientists – Silviculturist, Superintendent of Forest Working Plan, Forest Zoologist, Forest Botanist, Forest Chemist and Forest Economics. In 1921, the forests became a "transferred" subject and their administration vested with the Government of the concerned provinces. The realization concreted by November 1929 when IFRI moved to sylvan surround now famous as Forest Research Institute (FRI). Thereafter, rapid advances were made in forestry research, which valuably contributed towards the scientific management of state forests and development of forest based industries.

After independence, some other institutes of forestry research were established. In 1956, a regional research centre was established at Bangalore. Afterwards, three regional centres were established at Coimbatore in 1958, at Jabalpur in 1972 and at Jorhat in 1976. During the early 1970s, nearly 17 small centres of Forest-Vegetation Survey were established throughout the country.

FORESTRY RESEARCH AFTER INDEPENDENCE

In 1952, Govt. of India enunciated a new National Forest Policy of 1894 and making good the omissions noticed in it. It was based on the idea that each type of land should be allotted to that form of use under which it would produce most and deteriorate least. Forests were classified as Protection Forests, National Forests and Tree Lands. The policy prescribed that the country should aim at keeping out 1/3rd. of the geographical area under forest cover.

In 1970, National Commission on Agriculture was appointed which submitted its report in 1976 which included a full volume on forestry. The report highlighted the widening gaps between demand and supply positions of various forest products and pointed towards adopting a commercial approach for forestry development consistent with the environmental conservation. It emphasized the need for raising the per hectare production in volume and value creating more employment opportunities, giving a substantial support to the economy of backward areas and tribal population, expanding industries on forest based raw material, entering market in wood and wood products, and having a sustaining impact on the employment in the secondary and tertiary sectors.

It was around this period that substantial data base regarding forests was generated and trend for building up forest statistics was set in. Some relevant figures for this discussion are furnished below.

	Area in million hectares
1. India's geographical area	328.7263
2. Recorded forest area (1991)	77.0080 (23.426%)
3. Actual forest area (1991)	63.9182 (19.444%)
i) Dense forest (crown density 40% and above)	38.5008}
ii) Open forest (crown density 10% to less 40%)	24.9930} 63.9164
iii) Mangrove forest	0.4244}
4. Dense forest cover (crown density 40% and above)	
1971–73	46.42
1980–82	36.02
1986–87	35.77
1990–91	38.50
5. Forest area lost in million hectares for various purposes between	
a) 1951–52 to 1975–76	
i) River valley projects	0.4791
ii) Agricultural purposes	2,5069
iii) Construction of roads	0.0571
iv) Establishment of industries	0.1272
v) Miscellaneous purposes	0.9654
	4.1357
1976–77 to 1988–89	
@ 47500 ha/year	0.5700
Total (a + b)	4.7057
6. Forest area gained (1989–90 to 1990–91)	
@ 28000ha/year	0.0560
7. Land utilization in India	
i) Agriculture	154.70
ii) Forests	77.01
iii) Permanent pasture and other grazing lands	12.15
iv) Land under cultivable tree crops and groves	3.91
v) Cultivable waste land	16.64
vi) Land under other nonagricultural uses	17.53
vii) Barren and waste land	24.09
viii) Area for which no return exists	22.69
	328.72

A perusal of the above statistics shows that substantial forest area measuring 4.71 million hectares was lost during the period from 1951-52 up to 1989 in the name of development. It was not realized that maintenance of forest cover over the land is as important a developmental activity as the development in any other sector. It was perhaps due to the age old belief that the forests were in exhaustible and they recuperated naturally even though maltreated.

It is seen that vast chunks of land in the country need treatment in the form of afforestation, reforestation or tree culture, as follows:

24.9930 million hectares of open forests require rehabilitation. Out of the total forest land, 0.7 million hectares is under encroachment, 4.912 million hectares under shifting cultivation, 0.68 million hectares under mining, in addition to 16.64 million hectares cultivable waste lands, 24.09 million hectares barren and waste lands and other vacant lands are the potential areas which are in need of urgent treatment.

It is to be noted that some how or the other it has not been possible to go over all these areas and create national asset in the form of useful vegetation cover. Appropriate technology has also been lacking to tackle these difficult sites. It calls for serious and concerted research efforts.

It has been estimated that using various tree planting patterns, 2 to 5 % of farm land can grow trees for fuel and other purposes without any loss in agricultural production. Even if 2% of agricultural land (2% of 154.70 million hectares) is planted with trees, it will mean 3.1 million hectares of tree cover. For this purpose also, models suited to different areas will have to be evolved.

It has been stated earlier that the Govt. Forest Act 1865 took over the rights of communities and established state control over forests. The main aim of this enactment and subsequent legislations was to establish a total control over exploitation of forest resources.

The Indian Forest Act 1927 deliberated community rights and responsibilities. It was mentioned there "No person can claim a right to private properties in forest land merely because he has domiciled there, or even if his forefathers have lived there for centuries. Nor do such people have any rights over forest produce". The emphasis in this forest policy was on generating more revenue from forest areas.

In the post independence era, a policy which was framed in 1952 stated that in no way the rights of neighbouring communities on forest can be established in national interest. In case where villagers were involved in protecting and creating forests, they were granted right to distribute the produce only in case where the cost involved for such protection and creation comes from their own resources. Thus, the forest fringe villagers as per this law, were considered as only protectors and beneficiaries rather than partners and managers of forest resources even when they were closely associated in preserving forest resources.

The Forest Conservation Act 1980 imposed a total restriction on commercial logging. In spite of that during 1984, The National Remote Sensing Agency announced that between 1975 and 1982, India had lost 1.3 million hectares of forest annually. This failure to conserve in spite of very tight laws was explained by many as failure to " debureacaucratize" forest management and bring about the involvement of forest communities.

The present day effort of protecting forests by forest fringe people who are ecosystem people as well as ecological refugees do have antecedents in systems such as "Van Panchyats" of Kumaon, many of which have been functioning successfully since late 1920's. Results of CHIPKO movement in early 1970's inspired the idea of protecting forests by village people. In the Northern India, mostly in Tehri Garhwal Himalayas many Van Suraksha Samities (VSS) were formed by the local youth to protect their own forest resources from the external exploiters. Now-a-days, a large number of community forest management groups are functioning throughout India. These include the legally sanctioned Swayam Forest Groups of Kumaon, the Forest Protection Committees of West Bengal, the Hill Resources Management Societies of Haryana, the Village Forest Committees of Uttar Kannada, Karnataka and the informal indigenous groups operating in Orissa and Bihar.

The report of National Commission on Agriculture was an eye-opener for the Government. The effects of deforestation, which had already started showing in the form of loss of fertile soil, erroneous distribution of rainfall patterns, floods, droughts etc. on the one hand and reduced availability of timber, fuel wood, fodder and other forest produce on the other, started being recognized by one and all. As a result of this, serious deliberations at various levels in the Govt. took place which culminated in the formulation of the new National Forest Policy of 1988 (discussed later). Resource allocation to the forestry sector was increased manifold during the Seventh Plan. Side by side, the need to boost research support for scientific management of forest lands, which to an appreciable extent had got degraded or constrained due to overuse, was increasingly debated time and again.

BIOSPHERE RESERVE CONCEPT

At the initiative of UNESCO's Man and Biosphere (MAB) Programme, the Biosphere Reserve (BR) concept was initiated in 1974 for promoting *in-situ* conservation of pristine and representative ecosystems. Under National Biosphere Reserve Programme, India has created a network of protected areas in the form of 85 National Parks, 447 Wildlife Sanctuaries and 23 Tiger Reserves. The area covered under protected area network is 1,48,000 sq.km., which accounts for 4.5% of the total geographical area of the country. Broadly, following UNESCO's guidelines, the National Biosphere Reserve Programme was initiated in 1986 and is primarily aimed at:

- to serve as wider base for conservation of entire range of living resources and their ecological foundations in addition to already established protected area network system
- to bring out representative ecosystems under conservation and sustainable use on a long term basis
- to ensure participation of local inhabitants for effective management and to devise means of improving livelihood of the local inhabitants through sustainable use
- to integrate scientific research with traditional knowledge of conservation, education and training as a part of the overall management of BR.

In order to undertake activities relating to biodiversity conservation and development of sustainable management aspects, BR's are demarcated into 3 to 5 interrelated zones. These are i) natural or core zone, ii) manipulation or buffer zone, iii) transition or restoration zone and iv) stable cultural zone. In general, core zone and buffer zone are recognized for most of the BR's in India.

With the co-operation of State Governments/ Union Territories, following 14 Biosphere Reserves have been designated till March 2005.

Sl. No.	Name of the BR & total geographical area (sq.km)	Date of notification	Location (State)
1.	Nilgiri (5520)	01-08-86	Part of Wynad, Nagarhole, Bandipur and Madhumalai, Nilambur, Salient Valleys and Siruvani Hills (Tamil Nadu, Kerala and Karnataka
2.	Nanda Devi (2236.74)	18-01-88	Part of Chamoli, Pithoragarh and Bageshwar districts (UP)
3.	Nakrek (820)	01-09-88	Parts of Garo Hills (Meghalaya)
4.	Manas (2837)	14-03-89	Part of Kokrajhar, Bongaigaon, Barpeta, Nalbari, Kamprup and Darang districts (Assam)
5.	Sunderbans (9630)	29-03-89	Part of delta of Ganges and Brahamaputra river system (West Bengal)
6.	Gulf of Mannar (10500)	18-02-89	Indian part of Gulf of Mannar between India & Sri Lanka (Tamil Nadu)
7.	Great Nicobar (885)	06-01-89	Southern moist islands of Andaman and Nicobar
8.	Simlipal (4374)	21-06-94	Part of Mayurbhanj district (Orissa)
9.	Dibru-Saikhowa (765)	28-07-97	Part of Dibrugarhand Tinsukia districts (Assam)
10.	Dehang Debang (5111.5)	02-09-98	Part of Siang and Debang Valley of Arunachal Pradesh
11.	Pachmarhi (4926.28)	03-03-99	Part of Hoshangabad, Betul and Chhindwara district (Madhya Pradesh)
12.	Kangchendozonga (2619.92)	07-02-2000	Kangchendozononga National Park and surround park in Sikkim
13.	Agasthyamali (1701.00)	12-11-2001	Kerala
14.	Achanakmar-Amarkantak (3835.51)	30-03-2005	Part of Madhya Pradesh (1224.98) and part of Chhattisgarh (2610.53)

Following sites were identified by Indian MAB Committee of the Ministry of Environment and Forests, Govt. of India but yet to be designated as BR.

1. Namdapha Arunachal Pradesh
2. Thar Desert Rajasthan

3. Valley of Flowers Uttarakhand
4. Little Rann of Kutch Gujarat
5. Kaziranga Assam
6. Kanha Madhya Pradesh
7. North Islands of Andaman Andaman & Nicobar

20 Point Programme

In 1980, Government of India announced the revised 20-point programme in which "afforestation – tree planting" was incorporated as point No. 12 and higher priority was given to afforestation.

In 1985, National Wasteland Development Board was created to identify, afforest and ecologically regenerate wasteland with active participation of the people. The annual target for afforestation was kept at 5 million hectares.

During 1986, 20 point programme was revised and point No. 16 was included with new strategy for forest development as :

- Grow more trees and raise more forests with full involvement of local people
- Protect the traditional rights of tribal population and local communities of access to fuel and forest produce
- Reclaim wastelands for productive use
- Plant suitable and appropriate species to improve the vegetation in hills, deserts and coastal areas.

ENVIRONMENTAL STABILITY AND RESTORATION OF ECOLOGICAL BALANCE

The National Forest Policy of 1988 states in its preamble that over the years, forests in the country have suffered serious depletion; and that the depletion is attributable to relentless pressures arising from ever-increasing demand for fuel wood, fodder and timber; inadequacy of protection measures; diversion of forest land to non-forest uses without ensuring compensatory afforestation and essential environmental safeguards and a tendency to look upon forests as revenue earning resource. The document lays down the following objectives governing the policy:

- Maintenance of environmental stability through preservation and, where necessary restoration of the ecological balance that has been adversely disturbed by serious depletion of the forest of the country.
- Conserving the national heritage of the country by preserving the remaining natural forests with the vast variety of flora and fauna, which represent the remarkable biological diversity and genetic resources of the country.
- Checking soil erosion and denudation in the catchments areas of rivers, lakes, reservoirs in the interest of soil and water conservation for mitigating floods and droughts and for the retardation of silation of reservoirs.
- Checking the extension of sand-dunes in the desert areas of Rajasthan and along the coastal tracts.

- Increasing substantially the forest tree cover in the country through massive afforestation and social forestry programmes, especially on all denuded, degraded and unproductive lands.
- Meeting the requirements of fuel wood, fodder, minor forest produce and small timbers of the rural and tribal populations.
- Increasing the productivity of forests to meet essential national needs.
- Encouraging efficient utilization of forest produce and maximizing substitution of wood.

It states that the principal aim of policy must be to ensure environmental stability and the maintenance of ecological balance including the atmospheric equilibrium which are vital for sustenance of life forms, human, animal and plants. The deviation of direct economic benefits must be subordinated to this principal aim as essentials of forest management. It mentioned that:

- Existing forests and forest lands should be fully protected and their productivity improved. Forest and vegetal cover should be increased rapidly on hill slopes, in catchments of rivers, lakes and reservoirs and ocean shores and on semi-arid and desert tracts.
- Diversity of goods and productive agricultural lands to forestry should be discouraged in view of the need for increased food production.
- For the conservation of total biological diversity, the network of national parks, sanctuaries, biosphere reserves and other protected areas should be strengthened and extended adequately.
- Provision of sufficient fodder, fuel and pasture, specially in areas adjoining forests is necessary in order to prevent depletion of forests beyond the sustainable limit. Since fuel wood continues to be the predominant source of energy in rural areas, the programme of afforestation should be intensified with special emphasis on augmenting fuel wood production to meet the requirements of the rural people.
- Minor forest produce provides sustenance to tribal population and to other communities residing in and around the forests. Such produce should be protected, improved and their production enhanced with due regard to generation of employment and income.

To make it practicable the policy prescribes a minimum of 1/3rd of the total land area of the country to be under forest or tree cover. It suggests the following strategies with regard to forestry research, education and extension.

Forestry Research

- Boost scientific forestry research by strengthening research base and decide research priorities in the attainment of the policy goals.
- Increase forest productivity for higher yields of wood and other forest produce per unit area per unit time through application of modern scientific tools and technology.

- Develop a practicable technology to re-vegetate barren/marginal/waste/mined lands and rehabilitate water shades.
- Effectively conserve and manage existing forest resources (mainly natural forest eco-systems).
- Develop social forestry models compatible with land capabilities and resource needs of users, willingly adopted by rural/tribal folks as a contributor to their economy.
- Provide substitutes for food and wood products.

Forestry Education

- Provide recognition to forestry both as scientific discipline and as a profession.
- Formulate a meaningful curriculum and training model for forestry education at the Universities and other institutions to ensure professional competence and meaningful research endeavours for solutions to existing problems, as contributing in popularizing forestry as an economic activity in rural development.

Forestry Extension

- Encourage farmers to adopt silviculture alongside agriculture in various forms to provide for their daily needs and ensure self sufficiency in forest goods/services on a sustained basis.
- Bring home to farmers through short term extension courses and demonstration plots, the need for trees and their integration as a component of farming systems.
- Use mass media, audio-visual aids and extension literature in extending forestry practices and the message on the need for forests.

Establishment of ICFRE

While steps for formulation of the National Forest Policy were on, recognizing the importance of forestry research, education and extension, and their strengthening in the country, establishment of Indian Council of Forestry Research & Education (ICFRE) was announced in December 1986. It is to be noted that although the Imperial Forest research Institute, Dehra Dun, after its establishment in 1906, kept on developing and expanding from all angles and got recognition as one of the premier forestry research institutions of the world, it could not meet the vast research requirements of the nation. The objectives of the ICFRE are as follows:

1. To undertake, aid, promote and coordinate forestry education, research and its application.
2. To develop and maintain a national information centre for forestry and allied sciences.
3. To act as a clearing house of research and general information relating to forests and wild life.
4. To develop forestry extension programmes and propagate the same through mass-media, audio-visual aids and extension machinery.

5. To provide consultancy services in the field of research, education and training in forestry and allied subjects.
6. To do other things considered necessary to attain the above objectives.

This Council was granted autonomy in June 1991. The constituent institutes and centres under the Council are:

1. Forest Research Institute, Dehra Dun
2. Tropical Forest Research Institute, Jabalpur
3. Institute of Forest Genetics and Tree Breeding, Coimbatore
4. Institute of Wood Science and Technology, Bangalore
5. Rain Forest Research Institute, Jorhat
6. Arid Forest Research Institute, Jodhpur
7. Himalayan Forest Research Institute, Simla
8. Institute f Forest Productivity, Ranchi
9. Centre for Social Forestry & Eco-rehablitation, Allahabad
10. Forestry Research and Human Resource Development Centre, Chhindwara
11. Forest Research Centre, Hyderabad

The Council as well as the Institutes are in the process of development including creation of all necessary infrastructure to shoulder the research needs in forestry, in addition to forestry education and extension.

CHANGE IN FOREST COVER

All the exercise and efforts including legal, strengthening of research base, increasing funding in forestry sector, activities relating to rehabilitation of degraded forests, social forestry programmes, control over diversion of forest land for non-forestry purposes etc. have borne fruit. It is heartening to observe that during the years from 1989 to 1991, there has been an increase in the forest cover in the country at the rate of 28ooo hectares per year . The forest cover in the country according to 1999 assessment (FSI Report) was 637,293 sq.km. (dense forest : 377,358 sq.km., open forest : 255,064 sq.km., mangrove : 4871 sq.km.) while in 2001 it has been estimated as 675,538 sq.km. (dense forest : 416,809 sq.km., open forest : 258,729 sq.km.). Comparison of the two assessments shows an overall increase of 38,245 sq.km. or 6% in the forest cover in the country. There is a significant increase of 34,580 sq.km. or 9.0 per cent in the dense forest (mangrove cover included) while the open forest has increased by 3,665 sq.km. or 1.4 per cent.

Assessment year	Dense forest	Open forest	Total forest cover (sq.km.)
2001	416,809	258,029	675,538
2003	390,564	287,769	678,333
Change	—26,245	29,040	2,795

The net change in forest cover during the period between 2001 and 2003 assessments is estimated by comparing the extent of forest cover recorded in the two assessments. It is found that during this period, there has been a net increase of 2,795 sq.km. in overall forest cover.

The National Forest Policy (1988), aims at having a minimum of one third of the geographical area of the country under forest and tree cover and enjoins maintaining two third of the area in hills under forest cover in order to prevent erosion and land degradation and also to ensure maintenance of ecological balance and environmental stability> There are 123 districts in the country that can be classified as hill districts. The total forest cover in the hill districts of the country is 274,383 sq. km. constituting 38.77 per cent of the geographical area of these districts against the goal of 66 per cent. Out of total 123 hill districts, only 54 districts have more than 66 per cent forest cover (FSI Report) Of the rest, 36 hill districts have forest cover between 33 and 66 per cent and the remaining 33 districts have less than 33 per cent forest cover. Compared with 2001 assessment, the forest cover in hill districts has increased by 3,057sq.km. in 2003 (FSI Report).

GLOBAL CONSENSUS ON FORESTRY

Role of forests in national economy and maintenance of environment has not only been realized in India, but it has received recognition internationally. During 1990, UN General Assembly adopted a resolution on Environment and Development. It expressed major thrust on maintaining the quality of Earth's environment, sustainable development and combating deforestation, desertification and drought, and conservation of biological diversity.

Beijing Ministerial Declaration on environment and Development in June 1991 in which 14 countries participated, recommended multilateral measures to protect forests and promote their sustainable management, including the proposal for a global consensus on forestry, aiming at enhancing the economic, social and environmental potential of forests. It emphasized that the International Community, among other things, make efforts towards the greening of the world, and countries which had in the past destroyed their forest should increase their forest cover through reforestation and afforestation programmes.

The United Nations Conference on Environment and Development in the Earth Summit in Rio de Janeiro, Brazil during 3-14 June 1992 discussed global warming, climate change, climate variability, air pollution, ozen depletion, biodiversity conservation, and other related issues concerning the earth as a whole. It brought out that forest resources are essential to both development and the preservation of global environment. Using them rationally can create employment, help, alleviate poverty and provide a valuable range of products. It prescribes that to support and develop the multiple ecological, economic, social and cultural roles and functions of forests, trees and forest lands, the countries should strengthen forest related institutions and improve their technical and professional skills. Urgent action has been suggested to conserve existing forests and to expand areas under forest and tree cover. Sustainable development of forests has been emphasized. Other points brought home are carrying out re-vegetation and re-afforestation in all types of lands; developing planted forests to off set pressure on primary/old growth of forests, protecting forests from pollutants, mining and shifting cultivation, greening urban areas, acknowledging the role of forests as National Carbon Reservoirs and Sinks; application of genetics and biotechnology for improving productivity, the relationship between survival of forests and their contribution to human welfare should be incorporated into national economic accounting system Halting the spread of desert, protecting mountain ecosystem, meeting agricultural needs without destroying lands, sustaining biological diversity, environmentally sound management of biotechnology etc. have been adequately discussed and highlighted for action at international level.

Apart from the "Rio Declaration", the conference prepared a blue print for sustainable development to the year 2000 and beyond (Agenda 21), a non-legally binding authoritative statement of principles for a global consensus on the management, conservation and sustainable development of all types of forests (the "Forest Principles"). The principles provide for protection of ecologically viable representatives. With regard to access to biological resources, the sovereign right of the countries and the local communities where the forests are located is recognized.

In the above scenario, development of new participatory management systems by some of India's forest department and forest communities points towards a new promising direction in forest management, not just for India but for the rest of the world as well. The National Conservation Strategy on Environment and Development (1992) developed and adopted by the Govt. of India includes action to :

- Ensure sustainable and equitable use of resources for meeting the basic needs of the present and future generations without causing damage to the environment.
- Prevent and control future deterioration in land, water and air which constitute our life support systems.
- Restore ecologically degraded areas and strategies for environmental improvements in our rural and urban settlements.
- Prevent further damage to and conserve natural and man-made heritage.
- Ensure that developmental projects are correctly suited so as to minimize their adverse environmental consequences.
- Ensure that the environment and productivity of coastal areas and marine ecosystems are protected.
- Conserve and nurture the biological diversity, gene pool and other resources through environmentally sustainable development and management ecosystems with special emphasis on our montane, marine and coastal, desert, wetlands, riverine and island ecosystems.
- Protect the scenic landscape, areas of geo-morphological significance, unique and representative biomes and ecosystems and wild life habitat, heritage sites/ structures and areas of cultural heritage importance.

JOINT FOREST MANAGEMENT

The first promising step in the direction towards the development of new participatory management systems was taken by the Government of West Bengal through their orders of 12 July 1989 to constitute Forest Protection Committees (FPCs) who were allowed to take 25% of the timber produced as a measure of incentive. The order contained a detailed resolution on composition of FPCs, their duties, functions and usufructuary benefits to be provided to them. Some amendments were made to the resolution in July 1990 so as to make it more democratic and effective

In July 1990, the Govt. of India issued an order which stressed that the right holders should identify themselves with the development and protection of forests from which they benefits. The order asked the state forest departments to work out modalities to ensure participation of communities in afforestation programmes. It recommended that NGOs and voluntary agencies should act as interface between the forest communities for the revival, restoration and development of degraded forests. The institutions through which this could be achieved could be a Panchayat, a village co-operative or a village forest committee. The village community would get a share of usufructs, lops and tops and if protection has been successfully carried out, a share of the final revenue or crop.

A series of policies or orders supporting JFM by several state Governments have created new opportunities to respond to century-old conflicts between communities and state forest departments resulting rapid progress in establishing functional protection systems which have not only facilitated natural regeneration but also automatically increased forest productivity. JFM has spread over 14 Mha in 63,618 villages as on March 1, 2002. There is a need to enhance the rate of spread of JFM as well as performance with respect to participation of local communities. Clean Development Mechanism (CDM) of the Kyato Protocol provides an opportunity to promote JFM, to increase carbon stock in vegetation and soil, to build capacity and enable communities to participate effectively in forest protection and management and increase income or flow of benefits, creating long term stake for the communities.

FORESTRY AND CLIMATE CHANGE

Climate change is one of the most important global environmental problems facing humanity. Forest sector contributes about 20% of 7Gt C of carbon emissions annually. Forest sector also provides mitigation opportunity to the tune of about 2Gt C annually. Country commitments to reduce Green House Gas (GHG) are now being translated into corporate commitments to reduce GHG. Many corporations in developed countries are turning to corporations in developing countries and forming strategic alliances to meet their emission reduction commitments. The Kyato Protocol makes provision to take into account Afforestation, Reforestation and Deforestation and other agreed Land Use, Land Use Change and Forestry (LULUCF) in meeting their commitments. One of the ways suggested is to make project based activities aiming at reducing GHG or enhancing carbon stocks on a specified period through carbon sequestration. Such mechanism could offer potential benefits to farmers and pulp and paper industry by way of farm forestry/ industrial plantations initiatives as carbon offsets.

The discussion on mitigation in forest sector has raised several issues, which are the subjects of research, modeling and negotiations. The technical issues are very important and need to be addressed if forest sector has to play a critical role in mitigating climate change. In this context, an Workshop on "Forestry and Climate Change – Assessing Mitigation Potential and Costs" organized by ICFRE, Dehra Dun and Indian Institute of Science, Bangalore supported by U.S. Environmental Protection Agency (USEPA) and Lawrence Barkley National Laboratory, University of California, Barkley, USA on September

23 and 24, 2002 at New Delhi aims at addressing technical issues in forest sector with a focus on developing countries, particularly India. The Workshop also aims to identify key areas of research and capacity development and suggest future collaborative research and information sharing opportunities. The objectives of the Workshop are :

- Create awareness among different stakeholders on Climate Change – Forest Sector Mitigation and Project Opportunities.
- Discuss issues related to the GHG accounting of forest sector Projects : Performance, additionaiity, leakage, verification etc.
- Discuss methods, tools and techniques for assessing forest sector mitigation potential, costs.
- Understand the forest sector mitigation potential, costs and benefits in India.
- Learn from the experience of other developing countries in addressing Climate Change mitigation in forest sector (India, Indonesia, Philippines etc.)
- Discuss deforestation, afforestation, reforestation, additional activities and Project activities.
- Explore future areas for collaboration on research, forest sector case study development and methods for addressing key GHG accounting issues.

In a world where disastrous consequences of global warming are beginning to be felt, 2007 will be remembered as a year which brought the climate change issue on top of the world agenda, though some saw the efforts as still inadequate.

This was the year in which the world community for the first time gave a unified response to approaching reality of climate change caused by the concentration of GHGs in the atmosphere due to human activities, bringing in line the United States, the biggest emitter of these gases. The fourth assessment report of the UN Intergovernmental Panel on Climate Change (IPCC) brought out in the beginning of the year made it clear in no uncertain terms that time was running out for the planet unless drastic measures were taken to arrest the climate change by global warming.

It said that already enough amount of these gases had been emitted in atmosphere to cause a rise of around two degrees by 2050. It warned that global warming would lead to extreme droughts, floods, and rise in the sea level due to melting of polar ice threatening the existence of coastal communities, specially in poor countries.

The urgency of the report made around 200 nations in Bali, Italy towards the close of the year to chalk out a roadmap for talks to reach a treaty for succeeding the Kyato protocol under which developed countries have been given targets to cut emissions of GHGs.

CURRENT FORESTRY RESEARCH PRIORITIES

The value and importance of knowledge about forests is increasing and the kinds of information required differ those on only few years ago. Therefore, dramatic changes in forestry research and education are required. Changes and improvements should be made in the basic nature of forestry research, human resource devoted to forestry research, technology and information, transfer of results of forestry research and the funding of forestry research. Forestry research should be a continuing source of information from which improvements in national resource management can be made more sustainable and environmentally sound. Essentially, forestry research in India has to be conducted at two different levels : i) conservation level and ii) day-to-day management level. At the conservation, human intervention in the forest ecosystem is unwanted, whereas, at the management level human intervention is a must. The two levels may be contradictory but the policy-making national level research organizations have to find a balance between the two. Many issues such as biological diversity, impacts of pollutants on forest ecosystem, degraded land management must be addressed at very large spatial scales and over long period of time. Research at the scale of landscapes and regions will involve changes in the way forestry research is performed, including new mapping technology, collaborative with managers and user groups, creative experimental approaches and evaluation procedures that differ dramatically from those of forestry research. To meet these issues a large number of well educated scientists, technicians, extension specialists are needed The issues covering bilateral matching of institutions, sharing of benefits, specific sharing of management responsibilities and capacity building of institutions are to be standardized by the researchers.

Ministry of Environment and Forests (Govt. of India) has made Environmental Assessment mandatory for all major projects in which wildlife consideration are necessarily included as major part of assessment. It helps and ensures the protection of natural ecosystem and habitats of local species. In recent years, the Ministry has funded several projects on ecological research aimed at studies on protection of fragile ecosystem like mangroves, coral reefs and wetlands. The Ministry has already constituted a National Committee for the protection of these fragile ecosystems. This committee has identified 15 mangrove areas, 16 wetlands and 4 coral reefs for their conservation and management on priority. The Ministry has already provided financial assistance to several research projects undertaken in many Universities, research institutes (Trivedi 2007).

Some sincere efforts have been made the world over to conserve biological diversity and evolve strategies for sustainable development. The younger generation needs to be educated on the different aspects of working of nature and it is duty of the environmentalists/scientists/foresters to create an awareness among the general public and students in this regard. An international consensus on establishing global network for gene bank, microbial resource centres and marine parks is also important. Inventorization of plants and microbial diversity and *in-situ* and *ex-situ* conservation of natural resources are some of the important steps to be taken for conserving the biological diversity. At the same time, conservation must be coupled with socio-economic development especially in countries where population pressure threatens the natural biotic resources. Environmental pollution disturbs the food chain and food web in natural ecosystems. It bring unfavorable changes in its biotic and abiotic factors and threatens the existence of sensitive species. It eventually leads to destruction of rest of remaining flora and fauna due to changed climatic condition and associated effects. Most people recognize the urgent need for environmental education.

Biotechnology provides a range of tools and methods for assessment, monitoring and managing biological diversity such as clarifying taxonomic and evolutionary relationships among groups of organisms and assessing the effects of ecosystem disturbance or components of biological diversity and biological processes, Besides assessment, monitoring and managing, the tools of biotechnology could also be utilized for *in-situ* conservation i.e. assessment of optimal or minimal population size, and *ex-situ* conservation i.e. enhancing the quality of characteristics and efficiency through compact storage of DNA libraries and sequence data base (Apples *et al.* 1995). The tools of modern biotechnology are being increasingly applied for plant diversity characterization and undoubtedly they have a major role in assisting plant conservation programmes. The major areas of biotechnology which can assist plant conservation programmes are (Benson 2003) :

1. Molecular marker technology
2. Molecular diagnostics
3. Tissue culture (in vitro technology)
4. Cryopreservation
5. Information Technology

Sustainable NTFP (non-timber forest product) management is an important intervention to sustain community interest on the one hand and achievement of long term goal of sustainable forest management in India on the other. The conservation efforts do not only express itself at the species level but also at the soil level. Tropical forests have been traditionally used for timber products. Management for multiple uses, particularly for extraction of NTFPs poses a number of challenges because various species have different requirements for reproduction, recruitment and population growth. For many plant species, information on basic biology and silviculture is incomplete. Therefore, such studies have to be conducted to gather more and more information on NTFP management.

The increasing applications of cidal materials (e.g.) fungicides, herbicides, pesticides etc.) for higher food production, higher frequency of acid rains due to industrial emissions of oxides of sulphur and nitrogen, depleting ozen in the stratosphere due to chlorofluorocarbons and accumulation of xenobiotics in the habitats have already degraded diverse habitats and are responsible for a rapid loss of biodiversity. It is obvious that the pollutants lower the overall physiological activities of plant life by reducing chlorophyll, ascorbic acid, protein, carotenoid and increasing sugar and sulphate (Banerjee 2008). Some species have shown specific resistance to the prevalent pollutant concentration and have not shown much deviation in the biochemical parameters and some plants are highly sensitive. Biomonitoring using plants can be a simple and inexpensive process, which lends itself as potential, adaptable method of assessing air quality. The injury symptoms and biochemical changes of plants due to pollution provide reliable detection, recognition and monitoring of air pollutants. Sensitive plants, for future plant biomonitoring programme, may be used as indicator plants after standardization of artificially induced biochemical changes. On the other hand, the resistant species may be recommended for any afforestation programme to control the environmental pollution in the polluted areas. But there is a great need for an in-depth study in this direction.

Fly ash is the major solid waste product and also a particulate pollutant derived from the combustion of pulverized coal in thermal power plants. The pollutants, when released into the environment, affect vegetation directly or indirectly. With the recent introduction of the super thermal power plants, production and disposal of fly ash will continue to be one of the major national problems. A lot of emphasis is being laid on the biological and non-biological utilization of fly ash (Banerjee and Sett 2007). As the physicochemical properties of fly ash depends on the type and nature of coal burnt, its special effect with respect to the particular affected areas, specially on the vegetation is a matter of concern for the environment research explorers/foresters.

Mining is a vital sector of economy in almost all countries. But ecologically, mining is one of the most destructive activities of man. Majority of the mineral resources are associated with forests. Mining destroys the existing vegetation due to dumping of huge amount of overburden materials on adjacent un-mined land. Due to lack of organic matter and nutrients, the overburden materials are physically, chemically and biologically impoverished. Environmental impact and associated degradation with each stage of mineral exploitation is beyond the regenerative capacity of the ecosystem and has physical, chemical, biological, social and hydrological consequences.

Different types of mining have different types of impact, depending upon the nature of the mine and resources being exploited. Ecological restoration of mine spoils can be artificially initiated by providing some allogenic inputs like physical reclamation of spoils which enables more species to invade the spoil dumps . Process of succession can be accelerated by introducing herbaceous species particularly legumes and grasses other than tree species at the initial stage. Success in the establishment of plantations depends on selection of species suitable for the site, quality of planting stock, planting techniques appropriate to the climate and site, optimum use of proper amendments for maximum biomass production, use of proper mulch to conserve moisture etc. This is an area of research and planning. Although scientific research has made it feasible to reclaim a mined out site through afforestation , complete rehabilitation of such a site is still unseen. This is because of the fact that scientific reclamation of mined out site is a new activity in our country. The mined out site which has been afforested, needs continuous aftercare till the environment of the site or more correctly the environmental balance in the area, is considerably restored. How much time will be needed for this is still a matter of research.

Food and wood are two basic requirements of communities, which are provided by two major sectors, viz. Agriculture and Forestry. In a developing country like India, where about 80% of the population live in villages, people largely depend on their subsistence on the above two sectors. Agriculture undoubtedly plays the major role in their lives. Further, the interdependence of forestry and agriculture is an aspect to be given due weightage. Although urban communities have slightly sophisticated culture, still most people from the middle and lower income group depend on firewood for their fuel requirement. However, for the rural masses, it is the only source along with cow dung. Uncontrolled felling of forests for different purposes, and activities associated with development, have resulted in frequent floods and siltation in water reservoirs, erosion of top soil, nutrient losses, loss of biological diversity and reduction in sink area of green house gases etc.

Of the total land area nearly half are presently under agriculture and there is no chance of bringing additional area under crop production. The only scope for increasing the production of food grains, fuel and fodder is through the development of degraded lands and making their optimum use. With the introduction of agro-forestry system, it may be possible to solve the problems of fodder and fuel wood shortage and generate additional income, while conserving the productivity of the fields. Based on the information on the type of land, nature of the problems related to sustainable land use and adoptability of different tree species, suitable agro-forestry models can be developed. Greater emphasis should be given to utilize degraded lands for development of agro-forestry systems for conservation of natural resources viz. soil water and biodiversity, creating congenial and conducive microclimate for trees, under storey crops and micro-organisms.

FORESTRY RESEARCH IN ICFRE

Forestry research is the overall responsibility of the ICFRE. Besides the ICFRE, a number of Central and State level institutions (Botanical Survey of India, Zoological Survey of India, Forest Survey of India, Wildlife Institute of India, National Botanical Research Institute, National Bureau of Plant Genetic Resources, Indian Institute of Forest Management, Indian Grassland and Fodder Research Institute and a number of State Forest Research Institutes) are involved in carrying out forestry research. The ICFRE and its various research institutes and centres are engaged in continuation of ongoing forestry research programmes and to take further research in specific areas assigned to them as per their mandate.

With a view to conserve the forest resources and optimize their use for sustainable, the ICFRE has identified the following areas of priority research for 2000 A.D. and beyond :

1. Conservation, protection, regeneration , rehabilitation and sustainable development of natural forest ecosystem.
2. Enhancing productivity of wood and non-wood forest produce per unit area, per unit time by application of scientific and technological methods.
3. Ecological rehabilitation of all fragile ecosystems such as mountains, mangroves, deserts etc.
4. Revegetation of barren/marginal/waste/mined lands
5. Research on tree improvement
6. Research on improved utilization, recovery and processing of forest produce for value addition and employment generation.
7. Socio-economic and policy research for developing strategies towards attracting peoples participation in forest management and growing of non-wood forest produce and adoption of tree farming practices.
8. Conservation of biodiversity.

Activities of ICFRE and its research centres are in full swing on the above thrust areas to reduce pressure on natural resources by increasing productivity through genetic and silvicultural improvement and making available wood substitutes; eco-restoration through treatment of wastelands; conservation of forest ecosystems; tribal development and social forestry.

A comprehensive National Forestry Research Plan (NFRP) has been developed after working out due priorities looking to resource constraints. Regional research priorities were worked out in consultation with stakeholders by holding seminar/workshops in different States of the Union followed by a national workshop. For deciding projects and resource allotment, Research Advisory Committees have been constituted on which all the States Forest Departments have due representation and say.

With a view to increasing availability of high quality planting stock for afforestation/ reforestation purpose, funds were provided to State Government for establishment of seed production areas (SPAs), clonal seed orchards (CSOs), seedling seed orchards (SSOs) and vegetative multiplication garden (VMGs).

Implementation of modern nursery programme with root trainer is an important component of current research activities for the production of nursery stock and plant establishment and growth in the field.

There is tremendous scope for State Governments, forest based industries, unemployed youth and other user agencies to make use of environmental friendly technologies developed by ICFRE to generate wealth. These technologies will also go a long way in conserving forest resources and biodiversity of the country. At present, a large number of tested technologies are available for transfer. These are about use of secondary plantation species raised under agro-forestry/social forestry, value addition to products, improving forest productivity, new product/substitution of scarce products, and environmental conservation/reclamation. Technological packages for the rehabilitation of different mined overburdens have been developed and perfected. The time required for the natural eco-restoration of coal, copper, iron, manganese, iron and limestone mined areas has been worked out on the basis of physical, chemical, nutritional and biological characteristics of the overburden material of different mined areas. For developing effective sink area and green belt in industrially polluted areas, species have been identified, graded and indexed. A large number of pollutant resistant plant species have been identified which may be recommended for any afforestation programme to control the environmental pollution. Cost effective tissue culture protocol has been developed for *Dendrocalamus strictus* which is very difficult to root by conventional method of vegetative propagation. Locally fabricated low cost mist unit has been developed. For afforestation of water logged sodic site, suitable planting technique has been developed. Adhesives to be used in plywood industry were prepared from tannins of *Uncaria gambier* and *Acacia catechu.* A substitute of pectin from *Cassia tora* seed gum used in food industry was identified and a process of making this substitute was standardized.

National Forest Library and Information Centre (NFLIC) under ICFRE is the store house of knowledge and provides the same to State Forest Departments, Universities etc. through electronic net working . Grey literature on forestry with complete bibliographic details is being collected and computerized for facilitating speedy storage and retrieval of documents. Extension material like brochures, bulletins, annual reports, books, monographs, leaflets and video films is being produced and disseminated as a part of public awareness programme. ICFRE has launched its own Website globally accessible at http://www.icfre.up.nic.in.

The forestry research in India is poised for a big leap forward. There is much greater awareness at all levels in the country for forestry development and forestry research. The establishment of ICFRE to supervise, coordinate and advance forestry research is an outstanding development for forestry research in the country.

REFERENCES

Agrawal, V.P. 1985. *Forests in India, Environmental and Production Frontiers,* Oxford & IBH Publication Co., New Delhi.

Anonymous 1988. *National Forest Policy 1988,* Govt. of India, Ministry of Environment and Forests, New Delhi.

Anonymous 1991. *The State of Forest Report 1991,* Forest Survey of India, Ministry of Environment and Forest, Govt. of India, New Delhi.

Anonymous 2003. *The State of Forest Report, 2003,* Forest Survey of India, Ministry of Environment and Forest, Govt. of India, New Delhi.

Anonymous 1992. *Earth Summit – Agenda 21,* United Nation Conference on Environment & Development, Riode Janerio, Brazil, 3-14 June 1992.

Apples, R, Young, A.G., Tzotozos, G.T., Barlow, B.A. and Simoens, C. 1995. Biotechnology applications in biodiversity assessment and management. In : V.H. Heywood (ed.) ,*Global Biodiversity Assessment,* Cambridge University Press, Cambridge.

Banerjee, S.K. and Sett, R. 2007. Fly ash, characteristics, environmental impact and utilization. In : P.C. Trivedi (ed.), *Pollution Management,* Pointer Publishers, Jaipur, India.

Banerjee, S.K. 2008. Air pollution and vegetation. In : P.C. Trivedi (ed.), *Pollution and Bioremediation,* Aavishkar Publishers, Distributors, Jaipur, Rajasthan, India.

Benson, E.E. 2003. An introduction to plant conservation biotechnology. In : E.E. Benson (ed.), *Plant Conservation Biotechnology,* Taylor and Framis Group, printed in India by Replita Press Pvt. Ltd., Kundli.

Chakraborty, R. 1989. Forests – then and now – extent and classification – A study of M.P. with special reference to Mahakoshal Region. *J. Trop. Forestry,* **5** : 41-54.

Champion, H.G. 1936. A preliminary survey of the forest types of India and Burma. *Indian Forest Record (New Series),* **1** : 1-286.

Champion, H.G. and Seth, S.K. 1968. *Revised Survey of Forest Types of India,* Government of India Publication Division, New Delhi, pp. 404.

Lal, J.B. 1989. Status of India's forests, *J. Trop. Forestry,* **5** : 34-40.

Mcneely, J.A., Miller, K.R., Reid, W.E., Mittermieir, R.A. and Warner, T.B. (1990).*Conserving the World's Biodiversity IUCN Gland Switzerland,* WRI, C1, WWF US and World Bank, Washington D.C., USA.

Trivedi, P.C. 2007. *Global Biodiversity, Status and Conservation,* Pointer Publishers, Jaipur, Rajasthan, pp. 397.

❑❑❑

Chapter 15

Comparative Suitable Nutrient Solution for the Growth of Plants and its Effect on Calcium Deficiency

A.P. VERMA

Plant physiology is relatively a recent branch of botany in relation to such other branches as ecology, taxonomy, anatomy and morphology. Plant nutrition is probably one of the earliest of the many aspects of plant physiology to be studied quantitatively. Plant nutrition includes both organic and inorganic nutrition. The latter may also be called as mineral nutrition. In the subject of mineral nutrition, plant physiology bears a close relationship to agriculture. The practical importhance of the mineral nutrition of plants, through manurial practices or fertilizer programmes gave the emerging science of plant physiology much of its early impetus.

Nature has elaborated the form and substance of plants with surprisingly small number of elements of the periodic table. The elements of water and carbon dioxide (H,O and C) are also included in this category. These are usually treated separately. The remaining elements are distinctively of mineral origin. Even then, from their early origins in Aristotelian doctrine, the primitive concepts of mineral nutrition of plants made little progress through the middle ages. This was due to the various difficulties which will be enumerated later on.

Historically Van-Helmont (1577-1644), and later on others who resorted to experiments, ushered in the modern period in which the mineral nutrition of plants was to be based on the rational system of chemistry. He made what appears to have been the first quantitative attempt to gain some idea of the origin of the increase in the fresh weight of a willow branch. The experiment of Van-Helmont was reported by his son in 1684 in ortus Medicinae. Van-Helmont planted a willow *(Salix sp.)* cutting 5 lb in weight, in 200 lb of dry earth, and watered it with rain water over a period of 5 years. At the end

Department of Botany, Magadh University, A.N. College, Patna-800 013 India of the experiment the plant weighed 169 lb 3 oz, and the earth lost about 2 oz in dry weight. In other words plant gained about 164 lb at the cost of only 2 oz loss in dry weight of the soil. This experiment surely disproved Aristotle's view that plants absorb their food in an elaborated form from the soil, a view known as humus theory; but then Van-Helmont also made a mistake in concluding, perhaps understandably in the existing state of knowledge, that 164 lb of plant material had been produced from water alone. He did not explain the significance of the small decrease in dry weight of soil, which was presumably due to the removal of inorganic substances.

Woodward (1699) made the earliest recorded use of water culture method without any solid substrate, and grew vetch, potato and mint in water obtained from different sources. On the basis of his experiments he concluded that plants can survive without soil with their roots immersed in water. He found that growth in river water is better than that in rain water and best in a watery extract of soil. On the basis of these observation Woodward concluded that "earth, and not water, is the matter that constitutes vegetables". According to him water is mearly a carrier of "terrestrial matter", a view diagonally opposed to that of Van-Helmont. This work was followed sixty years later by experiments by a Frenchman, Duhamel Du Monceau (1758) who grew oak, almond, chestnut, walnut and beans in cultures with filtered water from the Seine. It goes to the credit of Hales (Vegetable staticks, 1727) who perceived for the first time that plants are nourished in part from air through the leaves. It was only firnly in the nineteenth century that a proper distinction could be made between Photosythesis and mineral nutrition.

De Saussure (1804), another Frenchman, was the first to use controlled water culture experiments. He grew *Polygonum persicaria* and *Bidens cannabina* in distilled water and in dilute solutions of various salts. He found (i) that nitrates are necessary in addition to other mineral salts; (ii) that the elements of water are fixed in the plants along with the carbon; (iii) that soil nitrogen and not the nitrogen of the air is used by the plants; (iv) and that the mineral solution taken up by the plants is extremely dilute. Surprisingly this work was extended only after a lapse of nearly forty years by Wiegman and Polstroff (1842). By means of refined techniques they concluded (i) that except for the mineral constituents present in the seeds, the rest of the mineral requirements are obtained by the plants by absorption through the roots; (ii) that the mineral substances of the plants are neither the products of life nor do they increase by life processes; (iii) and that all amounts above those contained in the seeds are acquired from the external medium. Gris (1844) recognised the importance of iron in plant growth.

According to Shive (1940) sand culture as a means of nutritional study was used first by Boussingault (1851-1856). The work of Boussaingault was further extended by Salm-Horstmar (1856) whereby he was able to omit one or more given elements from the nutrient solution. He proved that nitrogen, phosphorous, sulphur, calcium, magnesium, iron and manganese were necessary for the growth of oats. He also described visual deficiency symptoms including those of "grey speck" due to lack of manganese. The modern approach to the study of plant nutrition using sand culture methods was thus established by this time. The basis of modern water culture method rests on the later work of Sachs (1860) carried out at the school of forestry in Tharandt. According to Sachs water culture is simpler than the sand culture as it avoids the solid phase and the possible contamination when search is being made for mineral elements necessary for plant growth. The nutrient solution used by Sachs contained all those elements which were considered to be essential at that time. These elements were nitrogen, phosphorous, sulphur, potassium, calcium, magnesium and iron. In the same year knop (1860) standardised the methods for producing seedling material by germinating seeds on a muslin. Knop (1861 and 1865) by further experiments in varying the composition produced a nutrient solution simpler than that used by Sachs. This nutrient solution was based on molar ratios.

The firm establishment of water and sand culture methods have proved to be of the greatest value to the students of plant physiology in assessing nutritional requirements of plants. These methods have conclusively proved that plants can be grown successfully, generation after generation, in solution of pure chemicals. They have also disproved the claim that organic constituents are essential to the nutrition of green plants. By these techniques the nutritional requirements of the plants have been established in a effective manner. It would have been frankly impossible had it not been proved that the soil, a complex medium, is less essential to efficient culture. It was mainly due to the absence of these two techniques that the study of mineral nutrition of plants made little progress through the middle ages as mentioned earlier.

With the establishment of sand and water culture methods, refinement of technique and increased purity of chemicals, nutritional requirements of green plants have been generally established. According to Sutcliffe (1962) following elements are essential for the green plants. The table also shows the forms in which they are mainly absorbed. Non-ionic sources are indicated in brackets.

MACRONUTRIENTS

1.	Carbon	:	Bicarbonate, (Carbon-di-oxide).
2.	Hydrogen	:	Various anions, (Water).
3.	Oxygen	:	Various anions, (Water, gaseous oxygen).
4.	Nitrogen	:	Nitrate, Ammonium.
5.	Phosphorous	:	Phosphates.
6.	Sulphur	:	Sulphates.
7.	Potassium	:	Cations.
8.	Calcium	:	Cations.
9.	Magnesium	:	Cations.

MICRONUTRIENTS

1.	Boron	:	Borates.
2.	Chlorine	:	Chloride.
3.	Copper	:	Cations.
4.	Iron	:	Cations.

5. Manganese : Manganates.
6 Molybdenum : Molybdates.
7. Zinc : Cations.

Not only the essentiality of these elements has been established, the deficiency symptoms have also been studied. Most of these studies are concerned with flowering plants of economic uses. This will be quite clear from table I given by Hewitt (1952). One of the exceptions is *Lemna* which, though a flowering plant, is apparently of no economic importance. This plant has been extensively used for the study of its mineral requirements. The reasons for its extensive use in water culture experiments have been discussed by Steinberg (1941). For obvious reasons non-flowering green plants, which have no significant economic importance, have not attracted much attention of the students of plant nutrition. One of many such plants is *Salvinia* a free-floating, rootless, aquatic fern. There have been many physiological and ecological studies on *Salvinia.* One of these studies specifically deals with mineral nutrition of *Salvinia.* Blackman (1961) and Clatworthy and Harper (1962) grew this plant in non-sterile culture in an inorganic nutrient medium. They did not investigate the nutritional requirements of the plant. To be sure, Rao and Narayana (1968) studied the effect of minerals and auxins on *Salvinia auriculata* in aseptic cultures. However, their conclusions on mineral nutrition are of very general nature and deficiency symptoms are not well described. Gaudet and Koh (1968) also in the same year described the growth and development of *Salvinia* and its response to various growth ragulators in sterile culture. Again Gaudet (1973) studied the growth of *Salvinia* in sterile culture under standard conditions. In the last two studies *Salvinia,* no doubt, was grown in sterile inorganic nutrient solution, but no study has been done to find out either the essential role of the minerals or their deficiency symptoms. It was therefore decided to embark on a detail study of the mineral nutrition of *Salvinia* in order to find out the mineral elements essential for its growth and to find out the differences, if any, in the mineral requirements of this free floating, aquatic rootless fern (a non-flowering plant) and that of a flowering plant with roots. Attempt is also made to clearly establish the mineral deficiency symptoms. The present study is concerned only with the effect of calcium deficiency on the growth of *Salvinia* and its diagnostic symptoms. This study will form the basis of other similar studies on the mineral deficiency of elements one by one and thus an attempt will be made to get a complete picture of the mineral nutrition of *Salvinia*.

This work forms an integral part of a comprehensive project undertaken to study all the mineral nutritional requirements of *Salvinia*, and the effect of mineral deficiency of elements one by one, in the growth of this plant.

The choice of *Salvinia* in the present study has been governed by many considerations. First and foremost is the consideration, as mentioned in the introduction, that no detailed study of its mineral nutrition has been made. The plant is easily available in abundance from the garden pond. This is important since a large number of inoculums are required. In its aquatic habit, it is ecologically very well suited to water culture, the technique used in the present investigation. Being aquatic and rootless there is no special need for aeration of the nutrient solution. Since the plant is without root, there is also no need for darkening

Table : Comparative elemmental composition of the nutritive solutions c, e, f, g and h

Solutions	Total Concentration In Gm Per Liters	Ca	K	Mg	Na	N	S	P	Fe	O	H	C
C	1.9154	0.2291	0.0898	0.0599	-	0.1797	0.0962	0.0713	0.00055	1.14826	0.04275	0.00143
E	1.9154	-	0.4524	0.0599	-	0.1494	0.0962	0.0713	0.00055	1.04486	0.04275	0.00143
F	2.8534	0.2291	0.4524	0.0599	-	0.3095	0.0962	0.0713	0.00055	1.59356	0.04275	0.00143
G	2.8534	0.2291	0.0898	0.0599	-	0.3786	0.3237	0.0713	0.00055	1.60256	0.09995	0.00143
H	1.9154	-	0.0898	0.0599	-	0.2184	0.3237	0.0713	0.00055	1.05385	0.09995	0.00143

arrangement of the container. This arrangement is necessary in case of terrestrial plants growing in culture solutions since roots are negatively phototropic. Its rapid growth makes it a suitable material for growth analysis. The small size of the plant and absence of root would reduce the consumption of nutrient solution, a great advantage indeed. More over the surfaces of the floating leaves which form a large part of the plant body occupy a single horizontal surface. Thus there is no complication of self shading. In this way all the leaves are exposed to the same light intensity. In its aquatic, free-floating, autotrophic habit, in rapid vegetative multiplication, it duplicates *Lemna.* But while *Lemna* is a flowering plant with root, *Salvinia* is a rootless non-flowering plant. This study may thus provide a basis for comparing the nutritional requirements of an aquatic footless, non-flowering plant and an aquatic flowering plant with roots. These two plants resemble each other very closely in all other respects. Thus the growth behaviuor of *Salvinia* can be compared with *Lemna* on one hand and terrestrial flowering plants on the other.

According to Biswas and Calder (1937) there are only two species of *Salvinia, S. natans Hoffm.*and *S. cucullata roxb.* found in India. However, a Brazilian species *S. auriculate Aublet* has also been introduced and is flourishing quite well in Indian climate. *S. auriculata* is the species used in the present study which includes growth of *Salvinia* in a suitable culture solution. Afterwards the effect of the absence of calcium, a major nutrient, on its morphology and growth, and also the deficiency symptom will be studied. For the uniformity of the plant inoculum and to minimise sampling errors, length pieces of 2.5 cm. from the apex are taken with the leaves of approximately equal size and without axillary branches. This piece consists of three pairs of floating leaves and three submerged, dissected leaves. The inoculums were thoroughly washed in distilled water for removing any foreign body.

In the first place trial was made to find out a suitable nutrient solution. This is essential, since a large number of such solutions are reported in the literature, but none of them is uniformly and constantly suitable for all plant materials in different laboratory conditions.

To begin with , four different known culture solutions wee arbitrarily selected. These are Arnon and Hoagland nutrient solution. Hoagland and Snyder nutrient solution, Shive and Robbins nutrient solutions number I and ii as given by Hewitt (1952). These solutions, for convenience, are respectively labelled as A, B, C and D. The total concentration and elemental composition of the four nutrient solutions are given in Tables I, II, II and IV. Iron was added as tartarate and its concentration is given in respective tables. All the chemicals used except iton tartarate were analytical reagents. 500 ml of each of the four nutrient solutions were taken in different plastic containers since plastic material is chemically inert. Culture solutions were changed every third day during the experimental period of 28 days. This was done with a view to reduce marked fluctuation in the composition of nutrient solutions and in their P^{H}. Inoculumns were grown in each plastic container under a light intensity of 300 Lu/Sq.ft. for a daily period of 12-14 hours. The light intensity was measured by EEL photometer. For light source, a bank of four 2 ft. long day light fluorescent tubes of 20 watts each was used. The growth of the plants was determined in terms of the following growth parameters:

(a) Weekly increase in length;

(b) Weekly increase in fresh weight;

(c) Weekly increase in pairs of floating green leaves, and

(d) Total increase in dry weight.

For initial dry weight, dry weight of 8 or 16 standard size inoculums were taken and average determined. This average weight was taken to be the initial dry weight of a single inoculum. The dry weight was determined by drying the plants at 100^0c in an oven till a constant weight was obtained. All the data were statistically analysed to determine significant growth in any of the four solutions.

Of all the nutrient solutions in which observations were made, Shive and Robbins nutrient solution I (solution c) was found to give the best growth response. Therefore, in the study of the effect of calcium deficiency on the growth of *Salvinia* solution C was taken to be a normal solution, constituting a control , and a variant of it without calcium was used as a calcium deficient solution.

To study the effect of calcium deficiency on the internal structure of the different parts of the plant, they were fixed in F.A.A., dehydrated in different grades of ethyl alcohol and finally embedded in paraffine wax of melting point $52\text{-}54^0$ centrigrade in the usual way. Sections were cut on a rotary microtome at a thickness of 10-15/u and stained in Haematoxylin and orange "G". Details of fixing, dehydrating, embedding and staining procedures are as follows:

Fixative used was F.A.A. in which the constituents were mixed in the following proportion:

Formaldehyde	...	5 cc
Acetic acid	...	5 cc
70% Ethyl alcohol	...	90 cc

Material was kept for 12-14 hours in the fixative. After fixation, it was thoroughly washed in water, dehydrated in ethyl alcohol series and finally embedded in paraffin wax following, in general, the schedule of Johanson (1940). The details of the schedule are given below.

(1) Washing in water	...	3 changes.
(2) 30% alcohol	...	2 hours.
(3) 50% alcohol	...	2 hours.
(4) 70% alcohol	...	24 hours.
(5) 90% alcohol	...	1 hour.
(6) 100% alcohol	...	1 hour, two changes of 1/2hour each
(7) 75% of absolute alcohol plus 25% xylol	...	2 hours.
(8) 50% absolute alcohol		

plus 50% xylol. ... 2 hours.

(9) 25% absolute alcohol

plus 75% xylol. ... 2 hours.

(10) Xylol ... 2 changes of 2 hours each.

Paraffin wax then gradually added to the xylol containing the plant material which was finally embedded in wax. Blocks were prepared and sections were cut at 10-15/u. Ribbons were spread on albumenised glass slides. Details of the staining schedule are summerised below.

(1) Slides with ribbon placed in xylol to remove wax for 1 hour.

(2) Transferred to absolute alcohol and to descending series of alcohol for 10 minutes each.

(3) Washed with distilled water.

(4) Placed the section in 4% Iron alum for 10 minutes.

(5) Transferred to Haematoxylin for 30 minutes.

(6) Washed with distilled water.

(7) Placed section in 2% Iron alum for 5 minutes.

(8) Washed the section with distilled water.

(9) Placed sections in 30% alcohol for 5 minutes.

(10) Again placed the section in 50% alcohol for 5 minutes.

(11) Transferred the section in 70% alcohol for 5 minutes.

(12) Again transferred the section in 90% alcohol for 5 minutes.

(13) Placed the section in orange "G" for 1 ½ minutes.

(14) Transferred to absolute alcohol for 10 minutes.

(15) Transferred to xylol for 10 minutes.

(16) Mounted in canada balsam.

In the study of mineral nutrition of plants methods of sand and water culture have proved to be most valuable tool in the hands of the plant physiologists. *Salvinia,* being an aquatic plant can only be grown in water culture. In the study of its mineral nutrition, besides other things, two points are to be taken into consideration. First point is to have uniform inoculums and the second, to have a suitable nutrient solution. These two points are discussed one by one.

Uniform inoculums of *S. auriculata* were attempted by selecting randomly 2.5 cm length of the plant from apical side taking care that there are no lateral branches and the floating leaves are apparently of the same size. To find out if this method provides uniform inoculums, such pieces were subjected to fresh and dry weight determinations.

The results of the fresh and dry weight determinations indicate that if care is taken in selecting inoculums as stated above one can get fairly uniform inoculums. In the present investigation inoculums with three pairs of floating leaves were taken. Rao and Narayana (1968) have used rhizome pieces of *S. auriculata* with three pairs of dorsal leaves in their study on the effects of mineral nutrition and auxins. Rajan et al (1971) have used pieces of rhizome with a pair of fully expanded leaf and an apical bud in their study on the growth of *S. natans*. Again Gaudet (1973) used terminal pieces of various species of *Salvinia* consisting of two sets of leaves with two axillary and one terminal bud.

The next point to be discussed is the selection of suitable nutrient solution. A glance through the literature shows that a number of nutrient solutions have been used in sand and water culture methods. None of these is universally used because different nutrient solutions produce different growth response in different plant material. Not only that, different nutrient solutions have been used by different workers for the same plant material. To quote a few examples Bottomley (1920) grew *Lemna major* in Detemer and Knop's solution. *Lemna minor* was grown by Ashby et al (1928) in Clark's nutrient solution prepared in distilled water. Later on Ashby and Oxley (1935) grew *Lemna* again in Clark's nutrient solution but made in tap water. Gorham (1941) cultured *Lemna in* a nutrient solution different from those mentioned above. Hillaman (1959) grew *L. perpusilla* in Hutner's medium, Gorham's medium and L medium. Even in case of *Salvinia*, Clatworthy and Harper (1962) grew *S. natans* in non sterile Steinberg culture solution in which the concentration of potassium dihydrogen orthophosphate was 100 mg/L. Rajan et al (1971) on the other hand cultured *S. natans* in nonsterile Steinberg culture solution in which the concentration of potassium dihydrogen orthophosphate was four times than that of the original Steinberg culture solution. Gaudet and Koh (1968) cultured *S. rotundifolia* and Rao and Narayan (1968) grew *S. auriculata* in white's medium. Gaudet (1973) again grew different species of *Salvinia* in three different nutrient solutions. In view of these observations it was decided to find out a suitable nutrient solution for growing *S. auriculata* in water culture experiment under non sterile and ordinary laboratory conditions. For this four different nutrient solutions were selected. These are Arnon and Hoagland medium, Hoagland and Snyder medium, Shive and Robbins medium I and II referred to as solutions A, B, C, and D. Standard inoculums of the plant were grown in these solutions which were changed every third day to minimize changes in the composition of nutrient solution and p^H and also to reduce fungal infection, especially when the experiments are performed under non sterile conditions. Rajan et al (1971) also changed the culture solution in which they were growing. *S natans* every third day. No arrangement for aerating the nutrient solution was made since the plants are without root and are used to grow normally in water. It may be mentioned that Ashby and Oxley (1935) found that aeration of the nutrient solution was not necessary in case of *Lemna*, an aquatic rooted plant. There was also no need for arrangement for darkening the basal part of the plant, since roots which are negatively phototropic are absent in *Salvinia*.

The growth response of *S.auriculata* in terms of such growth parameters as increase in length, increase in fresh weight, increase in pair of floating fronds and increase in dry weight of the entire plant was determined in case of four nutrient solutions. From the perusal of the data obtained it can safely be concluded that solution C (Shive and Robbins nutrient solution I) consistently gave the best growth response. It may be mentioned here that Throughton (1967) also used Shive and Robbins nutrient solution

I in the study of growth in *L. perenne*, a rooted plant belonging to the family gramineae. In view of the observations made above, solution C has been used as a normal or control solution and a calcium deficient version of the solution C was prepared by substituting potassium nitrate for calcium nitrate. This calcium deficient solution is referred to as solution E. The solution C has an additional advantage. According to Smith and Butler (1921) a normal solution should not only produce a healthy growth but it could also be modified as to exclude calcium without any significant change occurring in concentration or elemental composition. Thus though solution E is without calcium, it does not very much differ from solution C.

It may be mentioned that according to Hillman (1961) growth measurements of Lemnaceae culture can be done by taking fresh or dry weights. According to Gorham (1950) as cited by Hillman dry weights measurements can be misleading because quite of ten they reflect primarily changes in starch content rather than growth. The most reliable measures of growth are determined by noting increase in frond number. In the present investigation instead of frond number weekly increase in pair of fronds has been used as one of the growth parameters. The result on fresh and dry weights, increase in pair of fronds and increase in length is consistent. Therefore so far *Salvinia* is concerned dry weight determination can be used as a good parameter of growth.

The determination of fresh and dry weights, frond number and length of *S. auriculata* in solution C and calcium deficient solution E clearly indicates better growth in solution C. Not only that, deficiency symptom on the floating fronds appear in calcium deficient solution E after about twelve days. A comparison of the elemental composition of solution C and E shows following differences. Solution E has no calcium, less nitrogen and more potassium. It is therefore desirable to check if the poor growth and deficiency symptoms produced by solution E is due to calcium deficiency or to other variables, such as less nitrogen, more potassium or change in P^H. The amount of nitrogen in solution E is not significantly different from that in solution C. This should not therefore influence the growth to a large extent. According to white (1937) the optomal nitrate concentration varies with the light intensity in case of *Lemna.* Nitrogen and decreases with decreasing light intensity requirement increases with increase in light intensity. In case of *S. auriculata* light intensity does not change. Thus nitrogen requirement should not significantly change. In any case a small fluctuation of 0.0303 mg/litre in nitrogen concentration should not be a significant factor if light intensity is kept constant.

The amount of potassium in solution E is however significantly much higher than that in solution C. But then it should normally increase growth rather than decrease it. According to Broyer and Stout (1959) increase in potassium concentration may possibly induce calcium and magnesium deficiency. This increase in potassium concentration in calcium deficient solution E might induce magnesium deficiency symptoms only, since calcium is already absent in solution E. That the deficiency symptoms are not due to magnesium deficiency, but due to calcium deficiency, are discussed further at a later stage.

So far the effect of P^H is concerned according to Rao and Narayana (1968) the best growth of *Salvinia auriculata* in modified White's medium is obtained at a P^H of 6.5 and change in P^H below and above the value does reduce the growth. Clatworthy and Harper (1962) grew *Salvinia natans* in non sterile Steinberg culture solution with a P^H of 5.2 to 5.3. Gaudet and Koh (1968) cultured *S. rotundifolia* in White's medium of P^H 4.5 Rajan et al.,(1971) grew *S. natans* in non sterile Steinberg culture solution with a P^H of 5.5. Gaudet (1973) grew different species of *Salvinia* in three different nutrient solutions but all having P^H 6.5.

Thus is appears that a change in P^H may be an important factor in producing growth response, but it may not be a very significant factor, in view of the fact that different workers have grown different species of *Salvinia* in different nutrient solutions with different P^H values. In this connection attention may be drawn to the work of Blackman and Robertson-Cuninghame (1953). They found that growth of *Lemna minor* in Steinberg solution was optimal at P^H 5.1, but did not show a large variation over a range of 4.6-6.1. But on the basis of deficiency symptoms obtained here and that reported in the literature for calcium deficiency it would appear that poor growth and deficiency symptoms produced in solution E are more likely due to absence of calcium than to any other variables. Even then this requires further investigation. As such a number of other solutions were prepared of which F and G are complete solutions and H a calcium deficient one. The total salt concentration and elemental composition of the solution C, E, F, G and H are given in text table . Of the complete solutions C, F and G used, solution C consistently give best growth response. The complete solutions F and G and calcium deficient solutions E and H produce poor growth in relation to solution C.

No deficiency symptom is obtained in solution F and G but deficiency symptom appear only in calcium deficient solution E and H. In solution F, increased potassium concentration does not induce any calcium or magnesium deficiency symptoms. Thus solutions E and H which are calcium deficient produce poorest growth and deficiency symptoms. It can therefore safely be concluded that poor growth and deficiency symptoms produced by solution E may be entirely due to calcium deficiency and not due to any other variables, such as total salt concentration or concentration of individual elements. This conclusion is further strengthened by the observation made by different workers on the effect of calcium deficiency on various plants. This aspect will be further discussed at a later stage. The effect of calcium deficiency on the growth of *S. auriculata* in terms of such growth parameters as length of plants, fresh and dry weights and pair of fronds has been studied. The results on all the four growth parameters are consistent. Increase in length, fresh and dry weights and pair of fronds is less in calcium deficient solution E in comparison to the normal solution C.

Poor growth of *S. auriculata* in terms of dry and fresh weights in calcium deficient solution has been reported by Rao and Narayana (1968). Recently Pichl, Ivan and Svatva Fialova (1971) also found retarded growth, a lower fresh and dry weights in case of Pumpkin and Pea when growing in calcium deficient solution. However according to them the wheat plant did not show these calcium deficiency symptoms; but the roots of wheat plants growing in calcium deficient solution did show a reduced fresh weight and significantly lower dry weight. Thus in its calcium requirements *Salvinia* is more akin to pumpkin and pea plants. The response of *Salvinia* to calcium deficiency is however similar to the root portion of the wheat plants. This finding of Pichl, Ivan and Svatva Fialova on the roots of the wheat plants is in conformity with the ovservation made earlier by Bamford (1931) on the roots of the wheat plants. Poor growth of the entire plant and the needle leaves of *Pinus taedia* in calcium deficient solution has also been reported by Davis (1949). On the other hand lettuce and onion according to Wallace (1951) show fairly normal growth even in absence of calcium. Tanaka and Woods (1973) have reported that calcium deficiency reduces root elongation in oat. Mamedova and Azizbekova (1973) have observed that growth of wheat is improved by increasing the concentration of calcium in the nutrient solution.

The decreased growth of *S. auriculata* is accompanied with visual deficiency symptoms. The symptoms are more evident in case of leaves. The symptoms start appearing after about twelve days. sorokin and Sommer (1929) found that plants of *Pisum sativum* growing in calcium deficient culture solutions usually die within two to three weeks. *S. auriculata* however continues to grow in absence of calcium even after third week, though the growth is very much reduced. Marrinos (1962) however found that calcium deficiency in Barley shoot apex became macroscopically evident within two days. This is probably due to the fact that the plants were transferred early to the calcium deficient solution when primary leaves had just emerged through the coleoptile tip and the remaining endosperm was also detached before transfer in order to remove residual seed nutrients.

As stated earlier the visual calcium deficiency symptom is more evident in case of floating leaves which is the major part of the plant. The first indication of deficiency symptom is obtained after about twelve days and it starts with the decay of the leaves from the margin resulting in a fringed margin of the older leaves. This is followed by hooking of the young leaf. This hooking of young leaf is a very distinct symptom. In the fourth week the entire plant breaks into pieces. Calcium deficiency symptom is also shown by the filiform segments of the ventral leaf. These filiform segments decay and break away resulting in a stubby appearance. The visual calcium deficiency symt- poms have been well illustrated by Wallace (1951) in his book. According to him outstanding symptoms of calcium deficiency usually appear in the young leaves and growing point of stem and roots. The leaves show sign of hooking. The margin of the leaf appears irregular in form and ragged. According to Wallace (1951) hooking of the leaf due to calcium deficiency is shown by Brassica crops, Broccoli, Brussels, Cabbage, Cauliflower, Kale, Rage, Sugar beat, Mangold and Seakable beat. Ragged and irregular leaf margin has also been reported for sugar beat, mango, seakale beat, potato and apple. The poor growth in calcium deficient solution is also accompanied with changes in the internal structure of the various parts of *S. auriculata*. The internal changes are reflected in the death and disorganisation of the thin walled cells. In case of the internode of the calcium deficient solution it has already been reported earlier that thin walled cells surrounding the stele are first to disintegrate. This tendency seems to extend from centre towards perifery. Also a comparison of the transverse section shows that the cross sectional area of the calcium fed internode is much more than that of the calcium starved internode. Albrecht and Davis (1929) have reported that the cell wall of the stem of the calcium starved seedling fail to retain their shape and apparently gave way before microtome blade, while the cells of the calcium fed plants stood up under the treatment and gave distinctly better sections. This may be due to the disorganisation of the cells. Low cross sectional area of young and old leaves with low cell number and cell size has been reported by Davis (1949) for *Pinus taedia* growing in calcium deficient solution. This also applied to the xylem and phloem. The primary tissue, specially cortex and pith in cross section of the young stem was also found to be much less. Thus smaller cross sectional area of calcium starved internode may be due to smaller size, smaller number of cells in pith and cortex and their partial disorganisation. This may be a plausible explanation in the light of the observations made by Lutman (1934) that calcium deficient stem in buck weed shows small sized epidermal cells, smaller vessels and small amount of pith. The observation of Lutman (1934) has been further reinforced by the findings of Lyon and Garcia (1944) that the amount of cortex, phloem, xylem and pith is less in calcium deficient stem than of the normal stem of tomato. Day (1929) on the other hand reported that calcium deficiency has no effect on the general structure of the mature root of *Pisum sativum*.

The floating leaves of calcium deficient plants also show disorganisation of the mesophyll tissue. This feature has also been noted by Wallace (1951) in brassica crops, broccoli, brussels, cabbage, cauliflower, kale and rape. The wilting of pedicels in runner bean and clover and of petiole in carrot may be due to disorganisation of inner cell due to calcium deficiency.

The collapse and disorganisation of the cells is understandable since calcium is an essential constituent of the middle lamella binding the cell walls of the different cells. This may result in distinct disorganisation of the delicate filliform segments of the ventral leaves resulting in stubby appearance. The reduction in growth in terms of length may be due to small size of the new cells or less number of new cells. In either case growth in terms of length will be less. Sorokin and Sommer (1929) while working on the root tips of *Pisum sativum* found that plants growing in calcium deficient culture solution usually die within two to three weeks. The death of the plants is associated with the degeneration in the protoplast of the meristematic cells of the root tips. The microscopic effect of calcium deficiency appears in the decreased amount of cytoplasm and in the appearance of large vacuole in the meristematic cells which therefore resemble old cells. This may result in decrease of the fresh and dry weights of calcium deficient *Salvinia auriculata*. According to Sorokin and Sommer (1929) the effect of calcium deficiency on the meristematic cells is due to one of the two reasons. Firstly calcium is a necessary constituent of the protoplast and therefore mitotic divisions are disturbed in its absence. Secondly calcium deficiency might so affect the physical condition of the colloidal system that normal mitosis is not possible. The effect of calcium deficiency is more marked in growing and young cells. If the growth of these cells is checked, growth of the entire plant is checked. Therefore, if calcium deficiency is to continue for a longer period, the plant may ultimately die. This is due to the fact that calcium is highly immoble and does not move from old to young part of the plant. This fact has been well emphasised by Gauch (1957). Ferell and Johnson (1956) have, however, reported that in western white pine, a tree species, calcium moves from old parts to newly developed buds. This may be an exceptional case.

According to White (1936,1938,1939) potassium deficiency decreases root length, multiplication rate, net assimilation and frond area in *Lemna minor*. Potassium deficiency also lowers root photosynthesis and reversibly increases respiration. Gradients with respect to plasmolysis time and permeablity to urea are also reduced. Pirson and Seidel (1950) and Pirson and Gollner (1953) again working with *L. minor* found that the effect of calcium deficiency is similar to the effect of potassium deficiency mentioned above. According to Hillman (1961) the sum evidence of the study of mineral nutrition of *L. minor* indicates that *Lemna* responds as a typical higher plant. Also it can be grown satisfactorily in light on a completely inorganic medium. The evidences of the present investigation also indicate that the rootless non-flowering aquatic *S. auriculata* resembles in a general way aquatic flowering *Lemna* with roots on one hand and higher plants on the other. Of course no direct reference has come to light which has described the visual calcium deficiency symptoms of *Lemna* and the effect that it produces on the internal structure.

REFERENCES

Albrecht, W.A. and Davis, F.L. 1929. Physiological importance of calcium in legume inoculation. *Bot. Gaz.* 88 : 310.

Ashby, E., Bolas, B.D. and Henderson, F.Y. 1928. The Interaction of Factors in the Growth of *Lemna*.I. Methods and Technique. *Ann. Bot.*, 42 :771.

and Oxley, T.A. 1935. The interaction of factors in the growth of *Lemna.* VI. An analysis of the influences of light intensity and temperature on the assimilation rate and the rate of frond multiplication. *Ann. Bot.*, 49: 309.

Bamford, R. 1931. changes in the root tips of wheat and corn grown in nutrient solutions deficient in calcium. *Bull. Torrey-Bot.Club.*, 58:149.

Biswas, K. and Calder, C.C. 1936. Hand book of common water and marsh plants of India and Burma.

Blackman, A. 1961. Responses to environmental factors by plants in the vegetative phase. In growth in living systems (Ed. Zarrow, et al). pp. 525. Basic Books, Inc. New York.

Blackman, G.E. and Robertson-Cuninghame, R.C. 1953. The influence of pH on the phytotoxicity of 2:4 di - chlorophenoxyacetic acid to *Lemna minor. New. Phytol.* 52: 71.

Bottomley, W.B. 1920. The growth *Lemna* plants in mineral solutions and their natural media. Ann. Bot. 34:345.

Boussingault, 1851-1856: cited by Shive,1940.

Broyer, T.C. and stout, P.R. 1959. The macronutrient elements. *Ann. Rev. Plant Physiol.* 10: 277.

Clatworthy, S. and Harper, J. 1962. the comparative biology of closely related species living in the same area. V. Inter and Intraspecific interference within cultures of *Lemna* spp. and *Salvinia natans. J.Exp. Bot.* 13:307.

Davis, D.E. 1949. Some effects of calcium deficiency on the anatomy of *Pinus Taedia. Amer. Jour. Bot.* 36: 276.

Day, D. 1929. Some effects of calcium deficiency on *Pisum sativum. Plant Physiol.* 4:493.

De Saussure, Th. 1804. "Researches chimiques sur la vegetation:". Nyon, Paris. 2.

Duhamel Du Monceau, H.L. 1758. "La Physique des Arbres" Guerin & Delatour (Paris). 2: 202.

Ferell, W.K. and Johnson, F.D. 1956. Mobility of Calcium-45 after injection into western white pine. *Science*, 124 : 364.

Gauch, H.G. 1957. Mineral nutrition of plants. *Ann. Rev. Plant Physiol.* 8 : 31.

Gaudet, J.J. 1973. Growth of a floating aquatic weed, *Salvinia*. Under standard conditions. *Hydrobiologia*, 41 :1 : : 77.

Gaudet, J.J. and Koh, D. 1968. Effect of various growth regulators on *Salvinia rotundifolia* in sterile culture. *Bull. Torrey Bot. Club.* 95 : 92.

Gorham, P.R. 1941. Measurement of the response of *Lemna* to growth promoting substances. *Amer. Jour. Bot.* 28 : 98.

Gorham, P.R. 1950. Heterotrophic nutrition of seed plants with particular reference to *Lemna minor*. *Canadian Jour. Res. C.*, 28 : 356.

Gris, E. 1844. Nourvelles experiences sur 1' action de composes ferugineux solubles applique a la vegetation et specialement en traitement de la chlorose et de la debilite des plantes. C. R. Acad. Sci. (Paris) 19 : 1118.

Hewitt, E.J. 1952. Sand and water culture methods used in the study of plants nutrition. Commonwealth Agricultural Bureaux, England.

Hillman, W. S. 1959. Experimental control of flowering in *Lemna* II. Some effects of medium composition, chelating agents and high temperature on flowering in *L. perpusilla.* 6746. *Amer. Jour. Bot.* 46 : 7 :: 489.

Hillman, W.S. 1961. The Lemnaceae, or Duckweeds arcview of the descriptive and experimental literature. *Bot. Rev.* 27 : 221.

Johansen, D.A. 1940. Plant micro technique. Mc Graw Hill Publication.

Knop, W. 1860. Uber die Ernahrung der pflanzendurch Wasserige Losungen bei Ausschluss des Bodens. Landw. *Verssta.* 2: 65, 270.

Knop, W. 1861. Quantitative analytische Arbeiten uber den Ernahrungsprocess der Pflansen. Landw. *Verssta.* 3 : 295.

Knop, W. 1865. Quantitative untersuchungen uber die Ernahrungsprocesse der pflanzen. Landw. *Verssta.* 7 : 93.

Lutman, B.F. 1934. Cell size and structure in plants as affected by various inorganic elements. Vermont *Agric. Exp. Sta. Bull.* 383 : 1.

Lyon, C.B. and Garcia, C.R. 1944. Anatomical responses of tomato stems to the macronutrient cation supply. *Bot. Gas.* 105 : 441.

Mamedova, S. Yu and Azizbekove, Z.S. 1973. Effect of various calcium doses on growth, development and translocation of mineral nutrients in wheat in case of carbonate salinization. IJV. Akad. Nauk. Azssrser. Bio. Nauk. 1:12.

Marinos, N.G. 1962. Studies on submicroscopic aspects of mineral deficiencies. I. Calcium deficiency in the shoot apex of Barley. *Amer. Jour. Bot.* 49 :8 ::834.

Pichl, Ivan and Svatva Fialova. 1971. Growth of wheat. pumpkin and pea in calcium and potassium deficiency. Acta. Univ. Carol. Biol. 2/3 : 135.

Pirson, A. and Seidel, F. 1950. Zell-und Stoffwechselphysiologische unterschungen an der Wurzel von *Lemna minor L.* unter besonderer Berucksichtigung von kalium-und kalziummangel. *planta*, 38 : 431.

Pirson, A. and Gollner, E. 1953. Zellphysiologische untersuchungen an der *Lemna* wurzel bei verminderter Nitrat-und Phosphatversorgung. *Zeits. Bot.* 41 : 141.

Rajan, A.K. Brenda Betteridge and Blsckman, G.E. 1971. Changes in the growth of *Salvina natans* induced by cycle of light and darkness of widely different duration. *Ann. Bot.* 35 : 597.

Rao, Y. and Narayana, R. 1968. Effects of mineral nutrition and auxine on *Salvinia auriculats* Aublet, grown in cultures. *J. of Mysore Univ.* 21 : 136.

Sachs, J.V. 1860. Berichte uber die physiologische thatigheit an der versuchsstation in Tharandt. IV vegetations versuche nut Ausschluss des Bodens uber die Nahrstoffe under sonstigen Ernahrungsbedingungen von Meis, Bohnen and anderen pflanzen. Landw. *Verssta.* 2 : 219.

Salm – Horstmar, F. 1856. Versuche und Resultate uber die Nahrung der pflanze Braunschwetg.

Shive, J.W. 1940. Sketch of the development of the water culture method of growing plants. *Sci. Mon.* 51 : 233.

Smith, T.O. and Butler, O. 1921. Relation of potassium to growth in plants. *Ann. Bot.* 35 : 189.

Sorokin, H. and Sommer, A.L. 1920. Changes in the cells and tissues of root tips. induced by absence of calcium. *Amer. Jour. Bot.* 16 : 23.

Steinberg, R.A. 1941. Use of *Lemna* for nutritional studies of green plants. *Jour. Agri. Res.* 62 : 423.

Sutcliffe, J.F. 1962. Mineral salts absorption in plants. Bergamon Press. Oxford, London – New York and Paris.

Tanaka, Yasuomi and Woods, F.W. 1973. Root and Roothair growth in oats. Replacibility of calcium. Canad. *Jour. Bot.* 51 : 1655.

Throughton, A. 1967. The effect of mineral on the distribution of growth in *Lolium perenne. Ann. Bot. n.s.* 31 :123 : :447.

*Van Helmont 1577-1644 ; cited in Ortus Medicinae, 1684.

Wallace, T. 1951. The diagnosis of mineral deficiencies in plants. A colour atlas and Guide. His Majesty's Stationary office, London.

*Wiegman and Polstroff 1842 – cited by Reed, page 242.

White, H.L. 1936. The interaction of factors in the growth of *Lemna*. VIII. The effect of potassium on growth and multiplication. *Ann. Bot.* 50 : 175.

White, H.L. 1937. The interaction of factors in the growth of *Lemna*. XI. The interaction of nitrogen and light intensity in relation to Growth and Assimilation. *Ann. Bot.* n.s. I : 4 ::623.

White, H.L. 1938. The interaction of factors in the growth of *Lemna* XIII. The interaction of potassium and light intensity in relation to root length. *Ann. Bot.* n. s. 2 : 911.

White, H.L. 1939. The interaction of factors in the growth of *Lemna* XIV. The interaction of potassium and light intensity in relation to growth and assimilation. *Ann. Bot.* n.s. 3: 619.

Woodward, J. 1699. Thoughts and experiments on vegetation. Phil. Trans. roy. Soc. (London). 21 : 382.

❑❑❑

Chapter 16

Microbial Endophytes : The Untapped Resources for Bioactive Extrolites

T. VIJAYA, K. CHANDRA MOULI AND K. RAMESH

In the continual search for new and useful compounds to provide assistance and relief in all aspects of the human condition is ever-growing. Natural products still remain the most important source for discovery of new and potential drug molecules. Bioactive extrolites from the endophytic microorganisms has been found superior in this regard. Microbial endophytes offer potential sources of novel natural products for exploitation in medicine, agriculture and the pharmaceutical industry (Strobel and Daisy, 2003). The term endophyte refers to interior colonization of plants by microorganisms that do not have pathogenic effects on their hosts (Reiter, 2002). In their symbiotic association, the host plant protects and feeds the endophyte, which "in return" produces bioactive metabolites to enhance the growth and competitiveness of the host and to protect it from herbivores and plant pathogens (Gunatilaka, 2006).

BIODIVERSITY OF ENDOPHYTES

All the vascular plant species examined to date were found to harbor endophytic bacteria and/or fungi (Arnold *et al.*, 2000). The colonization of endophytes in marine algae (Stanley, 1992) mosses and ferns (Raviraja and Sridhar, 1996) has also been recorded. Endophytes are important components of microbial biodiversity (Clay, 1992) commonly several to hundreds of endophyte species can be isolated from a single plant. The environmental conditions under which the host is growing also affect the endophyte population (Hata *et al.*, 1998) and the endophyte profile may be more diversified in tropical areas. Arnold *et al.*, (2000) isolated 418 endophyte species (estimated 347 genetically distinct taxa) from 83 healthy leaves of *Heisteria concinna* and *Ouratea lucens* in a lowland tropical forest and proposed that tropical endophytes themselves could be hyper diverse with host preference and spatial heterogeneity. Moreover, genotypic diversity has been

Department of Biotechnology, S.V.University, Tirupati-517 502, A.P, India.

observed in single endophyte species originating from conifers (Mc Cutcheon *et al.*, 1993) birch (Lappalainen and Yli-Mattila, 1999) and grasses (Reddy *et al.*, 1998).

Accordingly, endophytes are presumably ubiquitous in the plant kingdom with the population being dependent on host species and location. Endophytes are found in a wide variety of plant tissue types including seeds and ovules (Siegal and Latch, 1987), fruits (Schena *et al.*, 2003), stems (Gutierrez and Martinez, 2001), roots (Gemida *et al.*, 1998), leaves (Smith *et al.*, 1996), tubers (Sturz *et al.*, 1998), buds (Ragazzi *et al.*, 1999), xylem (Hoff *et al.*, 2004), rachis (Rodrigue and Samuls, 1999) and bark. (Raviraja, 2005). It is now widely accepted that endophyte-free plants are few and this is especially true for shrubs and trees (Gennaro *et al.*, 2003). Several studies have shown the presence of fungal endophytes in host species belonging virtually to all plant divisions, from mosses and ferns to monocotyledons. Literature suggests that both bacteria and fungi are the most common endophytic microorganisms (Petrini, 1986 and Petrini, 1999). Endophytic bacteria reside in plant tissues mainly in intercellular rather than intracellular spaces, and inside vascular tissues (Sessitsch *et al.*, 2002).

Endophytic microbes fall into several classes in relation to their plant organ source as Endophytes of dicots, monocots, lichens, bryophytes and ferns. The endophytes harbour tree bark, root, stem, leaves, galls and cysts.

Two major groups of endophytic fungi have been recognized previously, reflecting differences in evolutionary relatedness, taxonomy, plant hosts, and ecological functions: the clavicipitaceous endophytes (C-endophytes), which infect some grasses and the nonclavicipitaceous endophytes (NC-endophytes), which can be recovered from asymptomatic tissues of nonvascular plants, ferns and allies, conifers, and angiosperms. C-endophytes represent a small number of phylogenetically related clavicipitaceous species that are fastidious in culture and limited to some cool and warm season grasses (Bischoff & White, 2005). Typically these endophytes occur within plant shoots, where they form systemic intercellular infections. Clay & Schardl (2002) recognized three types of clavicipitaceous endophytes, ranging from symptomatic and pathogenic species (Type I) to mixed interaction and asymptomatic endophytes (Types II and III, respectively). NC-endophytes can be differentiated into three functional classes based on host colonization patterns, mechanism of transmission between host generations, *in planta* biodiversity levels, and ecological function.

Class 2 endophytes comprise a diversity of species, all of which are members of the Ascomycota or Basidiomycota. Most belong to the Ascomycota, with a minority of Basidiomycota. Members of the former are restricted to the Pezizomycotina, wherein they represent several classes. Class 2 endophytes within the Basidiomycota include a few members of the Agaricomycotina and Pucciniomycotina.

ISOLATION OF ENDOPHYTES

Endophytes live in close association with living plant tissues. It is known that perennial plants growing in tropical or semitropical areas of the world are hosts to a greater diversity of endophytes than those growing in drier or colder areas of the world having less plant diversity (Strobel, 2003). In order to obtain endophytes, should go for identification of plant species that may be of interest because of its unique biology, age, endemism, ethnobotanical history and environmental setting. This is an important step in the entire process of obtaining and studying endophytes, because it represents a biological basis in plant selection. Some endophyte groups may be undetected or under-represented and in particular, isolates that are unable to sporulate in culture may need to be evaluated by different approach. Host species, sampling strategy, host–endophyte and inter-endophyte interactions, tissue types and ages, geographic and habitat distributions, types of fungal colonisation, culture conditions, surface sterilants, and selective media all influence the detection and enumeration of endophytic fungi (Bills *et al.*, 2004). Bacon and White (1994) reviewed techniques used for isolation, maintenance, identification and preservation of grass endophytes. Schulz *et al.*, (1993) and Bills (1996) described detailed practical information on methods for the isolation of filamentous fungi from various substrata, including techniques, selective agents, and the most common media.

Strategies involved in the selection of plant for endophyte isolation and natural product. Several hypotheses govern this plant selection strategy, and these are as follows:

1. Plants from unique environmental settings, especially those with an unusual biology and possessing novel strategies for survival, are seriously considered for study.
2. Plants that have an ethnobotanical history (use by indigenous peoples) that is related to the specific uses or applications of interest are selected for study. These plants are chosen either by direct contact with local peoples or via local literature. Ultimately, it may be learned that the healing powers of the botanical source, in fact, may have nothing to do with the natural products of the plant, but of the endophyte (inhabiting the plant).
3. Plants that are endemic, having an unusual longevity or that have occupied a certain land mass are also more likely to lodge endophytes with active natural products than other plants.
4. Plants growing in areas of great biodiversity also have the prospect of housing endophytes with great biodiversity. (Strobel *et al.*, 2005)

Once the plant is selected for study, it is identified, and a small portion is harvested. Furthermore, it is to be exactly located geographically with a global positioning device. The small stem pieces are cut from the plant and placed in plastic bags after excess moisture are removed. Every attempt is to be made to store the materials at 4°C until isolation procedures can be initiated. Plant materials are thoroughly surface treated with 70% ethanol for 5 min and then treated with 0.2% mercuric chloride for 30 s until they dry under a laminar flow hood. Surface sterilisation of plant material is usually accomplished by treatment with a strong oxidant or general disinfectant for a period, followed by a sterile rinse. Household bleach (NaOCl), usually diluted in water to concentrations of 2–10%, is the most commonly used surface sterilant. Similarly effective oxidants include 3%H_2O_2 and 2%$KMnO_4$ (Bills *et al.*, 2004). Furthermore, the efficacy of surface sterilisation can be substantially improved by combining with a wetting agent.

This is particularly appropriate for hydrophobic or densely pubescent leaves. Ethanol (70–95%) is the most commonly used wetting agent in this respect, however, it has limited antibiotic activity, and thus is not used alone as a surface disinfectant (Schulz *et al.*, 1993). Other surfactants such as Tween 80 have found use as wetting agents to enhance surface sterilisation of the host plants (Bills *et al.*, 2004). The efficacy of sterilization was checked by rolling the leaves on 0.1% tryptic soy agar (TSA) plates. Then, with a sterile knife blade, outer tissues are removed from the samples and the inner tissues are carefully excised and placed on water agar plates. The leaves were homogenized with a sterilized mortar and pestle in phosphate-buffered saline.

To isolate endophytic microbes, a medium free of inorganic N was used, because the natural endophytic environment is very low in N (Tejera *et al.*, 2006). Different dilutions were placed on plates containing the semisolid N-free medium consisting of malic acid (5 g), K_2HPO_4 (0.5 g), $MgSO_4.7H_2O$ (0.2 g), NaCl (0.1 g), $CaCl_2$ (0.02 g), 0.5% bromothymol blue in 0.2 N KOH (2 ml), 1.64% Fe-EDTA solution (4 ml) and agar (2 g), per liter (Kirchhof *et al.*, 1997). The final pH was adjusted to 7.0 by KOH. To isolate pure cultures, bacterial isolates were sub cultured on TSA plates and incubated at 34 °C. Fungal isolates were subcultured on Sabouraud Dextrose Agar (SDA) plates and incubated at 27°C. Aspects of the biology and biochemistry of the endophytes are then examined (Strobel *et al.*, 1999). Pure cultures of endophytic bacteria are identified by Grams reaction, biochemical and molecular charecterization. Fungi were microscopically visualized under oil immersion lens by Lactophenol cotton blue (Collins *et al.*, 1989).

In order to isolate endophytes from plant seeds, the deglume is required to be removed together with contaminants associated with the dry glumes. This is achieved by rubbing the seeds vigorously between the hands and then rinsing the seeds for 15–20 minutes with a bleach solution (Bills *et al.*, 2004). The isolation of endophytic bacteria is often accomplished by pasting onto LB plates the trituration of plant tissues surface-disinfected with various disinfectants such as sodium hypochlorite, ethanol, hydrogen peroxide, mercuric chloride, or a combination of two or more (Hallmann *et al.*, 1997).

MORPHOLOGICAL AND MOLECULAR CHARACTERIZATION OF ENDOPHYTES

Morphological examination is performed by analyzing the culture, the mechanism of spore production, and the characteristics of the spores. This is especially valuable for isolates failing to produce spores or identifiable structures. Sometimes, optimisation of growth conditions aiming at inducing sporulation of endophytes is a trial-and-error process (Bills *et al.*, 2004). Each of the isolated fungal strains is separately inoculated on different media such as potato dextrose agar (PDA), sabouraud dextrose agar (SDA) media in petri dishes to achieve optimum conditions for sporulation (Reis *et al.*, 2000). Moreover, endophytic fungi that neither grow nor sporulate in culture can only be detected and identified by comparison of ribosomal DNA (rDNA) gene sequences that determine phylogenetic relationships (Guo *et al.*, 2000). Therefore, endophytic isolates are often identified using a combination of morphological and molecular methods (Malinowski and Belesky, 1999; Raps and Vidal, 1998). Concern has to be taken when closely related or morphologically similar endophytic fungi are under identification. The morphological features of some fungi are usually medium dependent and some cultural conditions can affect substantially vegetative and sexual incompatibility. Thus, the morphological character of endophytes should be coupled with the molecular evidence to enable significant isolation between closely related species.

For newly discovered endophytic fungi, morphology-based identification is confirmed by 18S rDNA sequence comparisons or internal transcribed spacer (ITS1 and ITS2) and 5.8S rDNA sequence examinations. Identification of fungal endophytes can best be accomplished by amplification of Internal Transcribed Spacer (ITS) regions, which are repeating units of DNA encoding ribosomal RNA. These regions evolve rapidly and there is a very large database of sequences on Genbank and the AFTOL (Assembly of the Fungal Tree of Life) project. Another sequence that is sometimes useful in identification of fungal species is the 18S ribosomal DNA. There is less diversity in this sequence across species, especially among families, but in conjunction with ITS sequence data, 18S rDNA data can be helpful (Nicholas, 2005).

Bacterial endophytic microbes are analysed by 16S rRNA base techniques. However, several molecular biological methods generally combined to identify unknown endophytic isolates. As an example, 16S rDNA cloning and sequencing, terminal restriction fragment length polymorphism (T-RFLP) analysis and denaturing or temperature gradient gel electrophoresis (D/TGGE) were combined to study diversity of bacterial endophytic populations in potato cultivars (Scott, 2001; Loiret *et al.*, 2004). Zuccaro *et al.* (2003) used DGGE with subsequent cloning and sequencing to identify the fungi associated with the macroalga *Fucus serratus*.

Despite the use of huge amounts of rDNA data available, some species within a given genus remain indistinguishable (Abbasi *et al.*, 2005). For that reason, new target sequences have been selected and validated. For instance, in order to distinguish some ten to eleven species, in addition to the rDNA fragment encompassing the internal transcribed spacers ITS1 and ITS2 and the 5.8 S sequence, cellobiohydrolase-C and the topoisomerase II genes, which were found to be single copy genes, were used as new molecular markers enhancing the accuracy of the analysis (Hatsch *et al.*, 2004). Furthermore, large subunit gene (LSU) of rDNA was a useful molecular marker for clarifying taxonomic and phylogenetic relationships in arbuscular mycorrhiza fungi (Silva *et al.*, 2006). Yet aiming better identification based on sequence analysis, programs as the oligonucleotide barcode TrichOKEY, and TrichoBLAST, a multilocus database of vouchered sequences, have been developed. In conjunction, these progresses enabled improved species classification (Druzhinina *et al.*, 2006).

Since many of the fungi isolated from plants are likely to be novel, it is helpful to have primers covering as many taxonomic groups as possible. Primers specific to bacteria can also be used to identify bacteria. Sequencing the amplified DNA and comparing it with the database can help determine whether or not the specimen is novel; if the sequence is known, the species can be identified (Nicholas, 2005).

Kowalchuk *et al.*, (1997) characterized 13 endophytic fungal isolates of marram grass (*Ammophila arenaria*) using primers for 18S rDNA. Primers for 28S rRNA were found to be more specific than those for 18S rRNA. Conversely, in attempts to amplify plant DNA using 'universal' primers for the ITS and 5.8S regions of rDNA, Zhang, Wendel & Clark (1997) also amplified fungal DNA from bamboos, as did Camacho *et al.*, (1997) from spruce needles, again emphasizing the importance of using specific primers.

Molecular techniques, as a powerful tool, are used to identify endophytic genera and species and explore the unique trophic niches occupied by non-sporulating and uncultivable fungi. However, the use of databanks, such as GenBank, for species identification is limited in several regards (Ragazzi *et al.*, 1999). It may be accepted as a trustworthy molecular technique once molecular databanks accurately reflect species collected from a broad range of geographic areas/environments.

BIOLOGICAL ROLES OF ENDOPHYTES

Endophytes colonizing inside plant tissues usually get nutrition and protection from the host plant. In return, they confer profoundly enhanced fitness to the host plants by producing certain functional metabolites. Endophyte-infected plants often grow faster than non-infected ones (Cheplick *et al.*, 1989). This effect is at least in part due to the endophytes' production of phytohormones such as indole-3-acetic acid (IAA), cytokines and other plant growth-promoting substances (Zou and Tan, 1999) and/or partly owing to the fact that endophytes could have enhanced the hosts' uptake of nutritional elements such as nitrogen (Reis *et al.*, 2000) and phosphorus (Gasoni and De Gurfinkel, 1997; Malinowski and Belesky, 1999) and that they regulate nutritional qualities such as the carbon–nitrogen ratio (Raps and Vidal, 1998). A culture broth of *Colletotrichum gloeosporioides*, an endophyte fungus of *Artemisia annua* L., has also been found to promote the growth of the host callus (Tan and Zou, 2001).

Certain endophytes improve the ecological adaptability of hosts by enhancing their tolerance to environmental stresses and resistance to phytopathogens and/or herbivores including some insects feeding on the host plant. Endophyte-infected grasses usually possess an increased tolerance to drought (Arachevaleta *et al.*, 1989; Ravel *et al.*, 1997) and aluminium toxicity (Malinowski and Belesky, 1999). Furthermore, some endophytes are able to provide the host plant with protection against some nematodes (Kimmons *et al.*, 1990; Hallmann and Sikora, 1996), mammal (Bacon *et al.*, 1977) and insect herbivores (Preszler *et al.*, 1996; Wilkinson *et al.*, 2000) as well as bacterial and fungal pathogens (Christensen, 1996; Sturz *et al.*, 1999). Some endophytes are capable of enhancing the hosts' allelopathic effects on other species co-growing nearby, usually being competitor (s) for the nutrition and the space (Sturz and Christie, 1996; Sturz *et al.*, 1998; Sutherland *et al.*, 1999). This could be the reason why some plants with special endophytes are usually competitive enough to become dominant species in successional fields (Clay and Holah, 1999).

BIOACTIVE EXTROLITES FROM ENDOPHYTES

Endophytes, microorganisms that reside in the tissues of living plants, are relatively unstudied and potential sources of novel natural products for exploitation in medicine, agriculture, and industry. It is noteworthy that, of the nearly 300,000 plant species that exist on the earth, each individual plant is host to one or more endophytes. Only a few these plants have ever been completely studied relative to their endophytic biology. The ingress to the human population of new diseases such as cardiac disease and severe acute respiratory syndrome requires the discovery and development of new drugs to combat them. Not only do diseases such as cardiac disease require drugs that target them specifically, but so do new therapies for treating ancillary infections which are a consequence of a weakened immune system. Furthermore, others who are immunocompromised (e.g., cancer and organ transplant patients) are at risk for opportunistic pathogens, such as *Aspergillus* spp., *Cryptococcus* spp., and *Candida* spp., that normally are not major problems in the human population. In addition, more drugs are needed to efficiently treat parasitic protozoan and nematodal infections, such as malaria, leishmaniasis, trypanomiasis, and filariasis.

Consequently, the opportunity to find new and interesting endophytic microorganisms among myriads of plants in different settings and ecosystems is great. As a general, a single endophytic strain will produce multiple bioactives (Strobel *et al.*, 1993). The reported natural products from endophytes include antibiotics, antipathogens, immunosupressants, anticancer compounds, antioxidant agents and other biologically bioactive extrolites.

ANTIBIOTICS FROM MICROBIAL ENDOPHYTES

Antibiotics are the most bioactive natural products isolated from endophytes. Strobel and Daisy have summarized the discovery of antibiotics from the first discovery of penicillin to most of the novel antibiotics isolated from endophytes up to 2003 (Strobel and Daisy, 2003). Many of them are proved to be of importance. New antimicrobial metabolites isolated from the culture of *Colletotrichum* sp. (Lu *et al.*, 2000). For example, *Artemisia annua* is a traditional Chinese herb that is well recognized for its synthesis of artemisinin (an antimalarial drug). The three new metabolites isolated from the culture of *Colletotrichum* sp. In *A.annua* were detected to not only have activity against human pathogenic fungi and bacteria but also be fungistatic to plant pathogenic fungi, reported that coronamycin, a complex of novel peptide antibiotics with activity against pythiaceous fungi and the human fungal pathogen *Cryptococcus neoformans*, produced by a verticillate *Streptomyces* sp. isolated as an endophyte from an epiphytic vine *Monstera* sp. It was also active against the malaria parasite, *Plasmodium falciparum*. Recently a *Streptomyces* sp. was reported on an annual plant–*Lolium perenne* (Guerney *et al.*, 1993). This lolium endophyte produces a weak antibiotic, designated methylalbonoursin, which is a diketopiperazine, condensed from leucine and phenylalanine. The streptomycete obtained from the snakevine plant (Castillo *et al.*, 2002). This microbe produced novel peptide antibiotics, designated munumbicins A, B, C, and D, in honor of Reggie Munumbi Miller. These antibiotics possess wide-spectrum activity against many human as well as plant pathogenic fungi and bacteria, and a *Plasmodium* sp. In general, the munumbicins demonstrated activity against Gram-positive bacteria such as *Bacillus anthracis*, multi-drug-resistant *M. tuberculosis*.

Cryptosporiopsis cf. quercina a fungus was isolated as an endophyte from *Tripterigeum wilfordii*, a medicinal plant native to Eurasia (Strobel *et al.*, 1999). A unique peptide antimycotic, termed cryptocandin, was isolated and characterized from *C. quercina* (Strobel *et al.*, 1999). This compound contains a number of peculiar hydroxylated amino acids and a novel amino acid-3-hydroxy-4-hydroxy methyl proline. The bioactive compound is related to the known antimycotics, the echinocandins and the pneumocandins (Walsh,

1992). Cryptocandin is also active against a number of plant pathogenic fungi, including *Sclerotinia sclerotiorum* and *Botrytis cinerea*. Cryptocin, a unique tetramic acid, is also produced by *C. quercina*. This unusual compound possesses potent activity against *Pyricularia oryzae*, the causal organism of one of the worst plant diseases in the world, as well as a number of other plant pathogenic fungi (Li *et al.*, 2000). Secondary metabolite is ambuic acid, an antifungal agent, which has been recently described from *P. microspora*, a rain forest endophytes (Li *et al.*, 2001).

A strain of *P. microspora* was also isolated from the endangered tree *Torreya taxifolia* and produces several compounds having antifungal activity including pestaloside, pyrones, pestalopyrone and hydroxypestalopyrone (Lee *et al.*, 1995).These products also possess phytotoxic properties. Other newly isolated secondary products obtained from *P. microspora* (endophytic on *Taxus brevifolia*) include two new caryophyllene sesquiterpenes, pestalotiopsins A and B (Pulici *et al.*, 1996a). Additional new sesquiterpenes produced by this fungus are 2R-hydroxydimeninol and a highly functionalized humulane (Pulici *et al.*, 1996b, Pulici *et al.*, 1996c). *Pestalotiopsis jester*, a endophytic fungal produces jesterone and hydroxyjesterone, which exhibit antifungal activity against a variety of plant pathogenic fungi (Li and Strobel, 2001). Phomopsichalasin, a metabolite from an endophytic *Phomopsis* sp., represents the first cytochalasin-type compound. This metabolite exhibits antibacterial activity against *Bacillus subtiils, Salmonella gallinarum,* and *Staphylococcus aureus*. It also displays a moderate activity against the yeast *Candida tropicalis* (Horn *et al.*, 1995). Colletotric acid, a metabolite of *Colletotrichum gloeosporioides*, an endophytic fungus isolated from *Artemisia mongolica*, displays antibacterial activity against bacteria as well as against the fungus *Helminthsporium sativum* (Zou *et al.*, 2000). Another *Colletotrichum* sp., isolated from *Artemisia annua*, produces bioactive metabolites that showed antimicrobial activity as well. *A. annua* is a traditional Chinese herb that is well recognized for its synthesis of artemisinin (an antimalarial drug). Some of *Pseudomonas* spp. produces phytotoxic compounds as well as antibiotics. The ecomycins are produced by *Pseudomonas viridiflava* (Miller *et al.*, 1998). The ecomycins are active against such human pathogenic fungi as *Cryptococcus neoformans* and *Candida albicans*. The pseudomycins produced by a plant-associated pseudomonads are another group of antifungal peptides (Harrison *et al.*, 1991; Ballio *et al.*, 1994). The pseudomycins are active against a variety of plant and human pathogenic fungi including *Candida albicans* and *Cryptococcus neoformans* and a variety of plant pathogenic fungi including *Ceratocystis ulmi* pathogen and *Mycosphaerella fijiensis* (causal agent of Black Sigatoka disease in bananas). Another endophytic *Streptomyces* sp. from *Grevillea pteridifolia* tree produces, in culture, novel antibiotics called kakadumycins that are related to the echinomycins. Echinomycin and kakadymycin A were studied as macromolecular synthesis inhibitors. Kakadumycin A significantly inhibited the RNA synthesis rate in *Bacillus subtilis* (Castillo *et al.*, 2003). Kakadumycin A also inhibited protein synthesis and cell wall synthesis substantially, but the effect was lower on DNA synthesis. Kakadumycin A shares a very similar inhibitory profile with echinomycin in four macromolecular synthesis assays. Kakadumycin A preferentially inhibits RNA synthesis and may have the same mode of action as echinomycin, which inhibits RNA synthesis by binding to a DNA template (Castillo *et al.*, 2003).

Endophyte	Plant source	Metabolite	Biological activity
Xylaria sp.	*Abies holophylla*	Griseofulvin Dechloro Griseofulvin	Antifungal antibiotic
Rhizoctonia sp.	*Cynodon dactylon*	Rhizoctonic acid	Anti-helicobacter pylori activity
Cytospora sp. *Diaporthe sp.*	Endophytic fungi from Guanacaste Conservation Area	Cytosporone	Antibacterial activity
Fusidium sp.		Fusidilactones	Antifungal activity
Curvularia lunata	*Niphates olemda*	Cytoskyrins	Antibacterial Activity
Curvularia lunata	*Niphates olemda*	Lunatin	Active against *Bacillus subtilis,* *Staphylococcus aureus,* *Escherichia coli,* *Cladosporium* *herbarum*
Streptomyces sp.	*Monstera sp.*	Coronamycin	Activity against pythiaceous fungi and the human fungal pathogen Cryptococcus neoformans, active against the malarial parasite–Plasmodium falciparum
Aspergillus *fumigates*	*Cynodon dactylon*	Asperfumoid, Physcion Asperfumin, Fumigaclavine C Fumitremorgin C	Inhibit growth of Candida albicans antifungal antifungal
Aspergillus niger	*Cynodon dactylon*	Rubrofusarin B Aurasperone A Fonsecinone A	Microbial pathogens antifungal
Penicillium sp.	*Melia azedarach*	Preaustinoid A, B	Exhibit moderate bacteriostatic effect on *Escherichia coli,* *Staphylococcus aureus,* *Pseudomonas*

Contd...

...Contd.

Endophyte	Plant source	Metabolite	Biological activity
			aeruginosa, Bacillus sp.
Streptomyces aureofaciens	*Zingiber officinale*	5,7-dimethoxy -4-phenylcoumarin 5,6-dimethoxy -4-(p-methoxyphenyl) coumarin Vanillin 3-methoxy-4-hydroxytoluene	antifungal
Serratia marcescens	*Rhyncholacis penicillata*	(-)-oocydin A	antifungal
Phomopsis sp.	*Erythrina crista-galli*	Phomol	antibacterial antifungal
Pestalotiopsis microspora	*Terminalia morobensis*	Pestacin Isopestacin	antimycotic antifungal
Pestalotiopsis spp.	several rain forest plants; leaf, stem, petiole	Ambuic acid	antimycotic
Periconia sp.	*Taxus cuspidata*	Periconicin A	antimycotic
Penicillium janczewskii	*Prumnopitys andina*	Peniprequinolone Mellein	nematicidal antibacterial antiviral
Nodulisporium sp.	*Bontia daphnoides*	Nodulisporic acid A Nodulisporic acid A1 Nodulisporic acid A2	insecticidal
Paecilomyces sp.	*Taxus mairei* *Torreya grandis*	Brefeldin A	antifungal; antiviral
Mycelia sterilia	*Atropa belladonna*	Preussomerin G-L	antibacterial antifungal
Muscodor albus	*Cinnamomum zeylanicum*	Volatile antibiotics	antibiotic
Hormonema sp	*Juniperus communis*	Enfumafungin	antifungal
Fusarium sp.	*Quercus variabilis*	Cerebroside Fusaruside	antibacterial antibacterial

Contd...

...Contd.

Endophyte	Plant source	Metabolite	Biological activity
Colletotrichum	*Ginkgo biloba* L.	Cytosporone	antifungal
Cytospora sp.	*Conocarpus erecta*		antibacterial
Diaporthe sp.	*Forsteronia spicata*	Cytosporone	antifungal antibacterial
Botrytis sp.	*Taxus brevifolia*	Ramulosin 6-hydroxyramulosin 8-dihydroramulosin	antibiotic
Aspergillus sp.	*Cynodon dactylon*	Monomethylsulochrin Helvolic acid	antibacterial antibacterial
Aspergillus niger	*Cynodon dactylon*	Fonsecinone A	antifungal
Aspergillus fumigates	*Cynodon dactylon*	Fumigaclavine C	antifungal
		Fumitremorgin C	antifungal

INHIBITORS FROM MICROBIAL ENDOPHYTES

Human cytomegalovirus (hCMV) protease inhibitors, cytonic acids A and B, have been isolated from the endophytic fungus *Cytonaema* sp (Guo *et al.*, 2000).

Endophyte	Plant source	Metabolite	Biological activity
Aspergillus niger	*Cynodon dactylon*	Rubrofusarin B Aurasperone A	Strong co inhibitors on xanthine oxidase
Aspergillus sp.	*Cynodon dactylon*	Monomethylsulochrin	Eosinophil inhibitor
Cladosporium herbarum	*Cynodon dactylon*	Aspernigrin A Rubrofusarin B	xanthine oxidase inhibitor
		Orlandin	plant growth inhibitor
Fusarium sp.	*Quercus variabilis*	Cerebroside	xanthine oxidase inhibitor
		Fusaruside	xanthine oxidase Inhibitor
Microsphaeropsis olivacea	*Pilgerodendron uviferum*	Graphislactone A	acetylcholinesterase (AChE) inhibitor
		Botrallin	acetylcholinesterase (AChE) inhibitor
Mycelia sterilia	*Atropa belladonna*	Preussomerin G	FPTase inhibitor

Contd...

...Contd.

Endophyte	Plant source	Metabolite	Biological activity
Nectria galligena	*Malus* x *domestica*	Colletochlorin B	acetylcholinesterase (AChE) inhibitor; β-glucuronidase inhibitor
		Ilicicolin C, E	AChE inhibitor β-glucuronidase inhibitor
		Ilicicolin F, α, β-dehydro curvularin	seed germination radicle and epicotyl growth inhibitor
Periconia sp.	*Taxus cuspidata*	Periconicin A, B	root growth inhibitor
Chaetomium chiversii	*Ephedra fasciculate*	Radicicol	Hsp90 inhibitor

ANTICANCER AGENTS FROM MICROBIAL ENDOPHYTES

Taxol and some of its derivatives represent the first major group of anticancer agents that are produced by endophytes. Taxol producing endophytic fungus, *Taxomyces andreanae,* was discovered in *Taxus brevifolia*. The original target diseases for this compound were ovarian and breast cancers, but now it is used to treat a number of other human tissue-proliferating diseases as well (Strobel *et al.,* 2004).

One of the most frequently isolated endophytic species is *Pestalotiopsis microspora*. An examination of the endophytes of *Taxus wallichiana* yielded *P. microspora,* and a preliminary monoclonal antibody test indicated that it might produce taxol (Strobel *et al.,* 1996).

Maguireothamnus speciosus, a rubiaceous plant, yielded a novel fungus, *Seimatoantlerium tepuiense* that produces taxol. Fungal taxol production has also been noted in *Periconia* sp (Li *et al.,* 1998) and *Seimatoantlerium nepalense,* another novel endophytic fungal species (Bashyal *et al.,* 1999). Furthermore, taxol is a fungicide, and the most sensitive organisms to it are plant pathogens such as *Pythium* spp. and *Phytophthora* spp (Young *et al.,* 1992).

Torreyanic acid, a selectively cytotoxic quinone dimer and potential anticancer agent, was isolated from a *P. microspora* strain. This strain was originally obtained as an endophyte associated with the endangered tree *Torreya taxifolia* (Florida torreya) (Lee *et al.,* 1996). Torreyanic acid was tested in several cancer cell lines, and it demonstrated 5-10 times more potent cytotoxicity in those lines that are sensitive to protein kinase C agonists and causes cell death by apoptosis.

Alkaloids are also commonly found in endophytic fungi. Such fungal genera as xylaria, phoma, hypoxylon, and chalara are representative producers of a relatively large group of substances known as the cytochalasins (Wagenaar *et al.*, 2000). Many of these compounds possess antitumor and antibiotic activities, but because of their cellular toxicity, they have not been developed into pharmaceuticals. Three novel cytochalasins have recently been reported from *Rhinocladiella* sp. as an endophyte on *Tripterygium wilfordii.*

Endophyte	Plant source	Metabolite	Biological activity
Pestalotiopsis microspora	*Torreya taxifolia*	Torreyanic acid	selectively cytotoxic activity
Aspergillus niger	*Cynodon dactylon*	Rubrofusarin B, Aurasperone A	inhibits colon cancer cells
Curvularia lunata	*Niphates olemda*	Cytoskyrins	potential anticancer agent
Apiospora montagnei	*Polysiphonia violacea*	Epiepoxydon significant	Exhibited cytoxicity against human cancer cell lines
Trametes hirsute	*Podophyllum hexandrum*	Podophyllotoxin, Aryl tetralin lignans	Anticancer activity
Fungal endophytic isolate	*Nothapodytes foetida*	Camptothecin	Anticancer activity
Aspergillus clavatus	*Taxus mairei* *Torreya grandis*	Brefeldin A	anticancer activity
Chaetomium chiversii	*Ephedra fasciculate*	Radicicol	cytotoxic
Fusarium oxysporum	*Catharanthus roseus*	Vincristine	anticancer
Paecilomyces sp.	*Taxus mairei* *Torreya grandis*	Brefeldin A	anticancer
Penicillium implicatum	*Diphylleia sinensis*	substance analogous to podophyllotoxin	anticancer
Colletotrichum	*Ginkgo biloba*	Cytosporone	cytotoxic
Cladosporium herbarum	*Cynodon dactylon*	Aspernigrin A	cytotoxic
Dothiorella sp.	*Aegiceras corniculatum*	Dithiorelone A, B Dithiorelone C, D	cytotoxic
Phoma medicaginis	*Medicago sativa* *Medicago lupulina*	Brefeldine A	initiation of apoptosis in cancer cells

ANTIOXIDANTS AND IMMUNOSUPPRESSIVE AGENTS FROM MICROBIAL ENDOPHYTES

Endophyte	Plant source	Metabolite	Biological activity
Pestalotiopsis microspora	*Terminalia morobensis*	Pestacin	antioxidant Isopestacin
Cephalosporium sp.	*Trachelospermum jasminoides*	Graphislactone A	antioxidant
Colletotrichum dematium	*Pteromischum sp.*	Collutellin A.	Immunosuppressive
Fusarium subglutinans	*Triptergium*	Subglutionol A, B *wilfordii*	Immunosuppressive

CONCLUSION

Endophytes are natural product synthesizers inside the plants. It is estimated that there may be as many as one million different endophytes species. Many of them are capable of synthesizing compounds that can be used by plants for defense against pathogens and some of the compounds have been proven as an outstanding source of both novel and bioactive natural products, which have an enormous potential for the development of new drugs and agricultural products. Endophytes are still out there to be found and developed and moreover they represent an abundant and dependable source of bioactive novel compounds for exploitation.

The recent development in the field of microbial natural products is the demonstration of the potential to bring up a variety of new metabolites from a single strain of a microorganism by systematic alteration of its cultivation parameters, known as OSMAC (one strain many compounds) approach. Such approaches should be exploited in the direction of finding new drugs that may be effective candidates for treating newly developing diseases in humans, plants, and animal.

REFERENCES

Abbasi, M., Goodwin, S. B. and Scholler, M. 2005. *Mycoscience*, 46, 241-247.

Arachevaleta, M., Bacon, C. W., Hoveland, C. S. and Radcliffe, D. E. 1989. *Agron. J.*, 81, 83.

Arnold, A. E., Maynard, Z., Gilbert, G. S., Coley, P. D. and Kursar, T. A. 2000. *Ecol. Lett.*, 3, 267.

Bacon, C. W. and White, J. F. in *Biotechnology of Endophytic Fungi of Grasses*, ed. C.

Bacon, W. and White, J. F. 1994. CRC Press, Boca Raton, FL, USA, pp. 47–56.

Bacon, W., Porter, J. K., Robbins, J. D. and Luttrell, E. S. 1977. *Appl. Environ. Microbiol.*, 34, 576.

Ballio, A., Bossa, F., DiGiogio, P., Ferranti, P., Paci, M., Pucci, P., Scaloni, A., Segre, A., Strobel, G. A. 1994. *FEBS Lett.* 355, 96-100.

Bashyal, B., Li, J. Y., Strobel, G. A., Hess, W. M. 1999. *Mycotaxon* 72, 33-42.

Bills, G. F. 1996. "Isolation and analysis of endophytic fungal communities from woody plants' in *Endophytic Fungi in Grasses and Woody Plants: Systematics, Ecology and Evolution*, ed. S. C. Redlin and L. M. Carris, American Phytopathological Society Press, St. Paul, MN, USA, pp. 31–65.

Bills, J. K., Christensen, M., Powell, M. and Thorn, G. 2004. in *Biodiversity of Fungi: Endophytic Fungi*, ed. G. M. Mueller, G. F. Bills and M. Foster, Elsevier Academic Press, CA, USA, p. 241.

Bischoff, J.F., White, J.F. Jr 2005. Evolutionary development of the Clavicipitaceae.

Camacho, F. J., Gernandt, D. S., Liston, A., Stone, J. K. and Klein, A. S. 1997. *Molecular Ecology*, 6, 983–987.

Castillo, U., Harper, J. K., Strobel, G. A., Sears, J., Alesi, K., Ford, E., Lin, J., Hunter, M., Maranta, M., Ge, H., Yaver, D., Jensen, J. B., Porter, H., Robison, R., Millar, D., Hess, W. M., Condron, M., Teplow, D. 2003. *FEMS Lett. 224*, 183-190.

Castillo, U., Strobel, G.A., Ford, E.J., Hess, W.M., Porter, H., Jensen, J.B., Albert, H., Robison, R., Condron, M.A., Teplow, D.B., Stevens, D., Yaver, D. 2002. Munumbicins, wide spectrum antibiotics produced by *Streptomyces munumbi*, endophytic on *Kennedia nigriscans*, Microbiol, 148, 2675–2685.

Cheplick, G. P., Clay, K. and Marks, S. 1989. *New Phytol.*, 111, 89.

Christensen, M. J. 1996. *Aust. Plant Pathol.*, 25, 186.

Clay, K. 1992. *Nat. Toxins*, 1, 47.

Clay, K. and Holah, J. 1999. *Science*, 285, 1742.

Collins, C.H., Lyne, P. M. and Grange, J. M. 1989. *Collins and Lyne's microbial methods*, 6th edition (London: Butter-worths) p. 95

Cutcheon, T. L. Mc, Carroll, G. C. and Schwab, S. 1993. *Mycologia*, 85, 180.

Dighton, J., White, J.F., Oudemans, P., eds. *The fungal community: Its organization and role in the ecosystem*. Boca Raton, FL, USA: Taylor & Francis, 505–518.

Druzhinina, I. S., Kopchinskiy, A. G. and Kubicek, C. P. 2006. *Mycoscience*, 47, 55-64.

Gasoni, L. and Gurfinkel, B. S. De 1997. *Mycol. Res.*, 101, 867.

Gennaro, M., Gonthier, P. and Nicolotti, G. 2003. *J. Phytopathol.*, 151, 529.

Germida, J. J., Siciliano, S. D., Freitas, R. de and Seib, A. M. 1998. *FEMS Microbiol. Ecol.*, 26, 43.

Guerney, K.A., Mantle, P.G. 1993. Biosynthesis of 1-*N*-methylalbonoursin by an endophytic *Streptomyces* sp, J. Nat. Prod 56, 1194–1198.

Gunatilaka, A. A. L. 2006. *J. Nat. Prod.* 69:509-526.

Guo, B., Dai, J., Ng, S., Huang, Y., Leong, C., Ong, W., Carte, B. K. 2000. *J. Nat. Prod.* 63, 602-604.

Guo, L.D., Hyde, K.D. and ELiew, C. Y. 2000. *New Phytol.*, 147, 617.

Gutierrez-Zmora, M.L. and Martinez-Romero, E. 2001. *J. Biotechnol.*, 91, 117.

Hallmann, J., QuadtHallmann, A., Mahaffee, W. F. and Kloepper, J.W. 1997. *Can. J. Microbiol.*, 43, 895.

Hallmannm, J. and Sikora, R. A. 1996. *Eur. J. Plant Pathol.*, 102, 155.

Harrison, L., Teplow, D., Rinaldi, M., Strobel, G. A. 1991. *J. Gen. Microbiol.* 137, 2857-2865.

Hata, K., Futai K. and Tsuda, M. 1998. *Can. J. Bot.*, 76, 245.

Hatsch, D., Phalip, V. and Jeltsch, J. 2004. *Res. Microbiol*, 155, 290-296.

Hoff, J. A., Klopfenstein, N. B., McDonald, G.I., Tonn, J. R., Kim, M. S., Zambino, P. J., Hessburg, P. F., Rogers, J. D., Peever, T. L. and Carris, L. M. 2004. *For. Pathol.*, 34, 255.

Horn, W. S., Simmonds, M. S. J., Schwartz, R. E., Blaney, W. M. 1995. *Tetrahedron*, 14, 3969-3978.

Kirchhof, G., Reis, V. M., Baldani, J. I., Eckart, B., Dobereiner, J. and Hartman, A. 1997. *Plant Soil*, 194, 45–56

Kowalchuk, G., Gerards, S. and Woldendorp, J. W. 1997. *Applied and Environmental Microbiology*, 63, 3858–3865.

Lappalainen, J. H. and Yli-Mattila, T. 1999. *Myocl. Res.*, 103, 328.

Lee, J. C, Strobel, G. A., Lobkovsky, E, Clardy, J. C. 1996. *J. Org. Chem.* 61, 3232-3233.

Lee, J. C, Yang, X, Schwartz, M, Strobel, G. A, Clardy, J. 1995. *Chem. Biol.* 2, 721-727.

Li, J. Y, Harper, J. K, Grant, D. M, Tombe, B. O, Bashyal, B, Hess, W. M, Strobel, G. A. 2001. *Phytochemistry 56*, 463-468.

Li, J. Y, Sidhu, R. S, Ford, E, Hess, W. M, Strobel, G. A. 1998. *J. Ind. Microbiol, 20*, 259-264.

Li, J. Y, Strobel, G. A, Harper, J. K. 2000. Lobkovsky, E, Clardy, J. *Org. Lett, 2*, 767-770.

Li, J. Y.; Strobel, G. A. 2001. *Phytochemistry, 57*, 261-265.

Loiret, F. G., Ortega, E., Kleiner, D., Ortega-Rodes, P., Rodes, R. and Dong, Z. 2004. *J. Appl. Microbiol.*, 97, 504.

Lu, H., Zou, W.X., Meng, J.C., Hu, J., and Tan, R.X. 2000. *Plant Sci.*, vol. 151, pp. 67–73.

Malinowski, D. P. and Belesky, D. P. 1999. *J. Plant Nutr.*, 22, 835.

Miller, R. V.; Miller, C. M.; Garton-Kinney, D.; Redgrave, G. B.; Sears, J.; Condron, M.; Teplow, D.; Strobel, G. A. 1998. *J. Appl. Microbiol. 84*, 937-944.

Nicholas, J. Hundley 2005. Structure elucidation of bioactive compounds isolated from endophytes of Alstonia scholaris and Acmena graveolens. Thesis submitted to the Brigham Young University (Master of Science), Department of chemistry and biochemistry. pp. 13-14.

Petrini, O. 1986. in *Microbiology of the Phyllosphere*, ed. J. H. Andrews and J. van den Heuvel, Cambridge University Press, Cambridge, p. 392.

Petrini, O. 1991. *Microbial Ecology of Leaves*, ed. J. H. Andrews and S. S. Hirano, Springer, New York, p. 499.

Preszler, R. W., Gaylord, E. S. and Boecklen, W. J. 1996. *Oecologia*, 108, 159.

Pulici, M., Sugawara, F., Koshino, H., Uzawa, J., Yoshida, S., Lobkovsky, E., Clardy, J. 1996. *J. Org. Chem, 61*, 2122-2124.

Pulici, M., Sugawara, F., Koshino, H., Uzawa, J., Yoshida, S., Lobkovsky, E., Clardy, J. 1996. *J. Nat. Prod, 59*, 47-48.

Pulici, M., Sugawara, F., Koshino, H., Uzawa, J., Yoshida, S., Lobkovsky, E., Clardy, J. 1996. *J. Chem. Res*, 378-379.

Ragazzi, A., Moricca, S., Capretti, P. and Dellavalle, I. 1999. *J. Phytopathol.*, 147, 437.

Raps, A. and Vidal, S. 1998. *Oecologia*, 114, 541.

Ravel, F., Courty, C., Coudret, A. and Charmet, G. 1997. *Agronomie (Paris)*, 17, 173.

Raviraja, N. S. 2005. *J. Basic Microbiol.*, 45, 230.

Raviraja, N. S., Sridhar, K. R. and Barlocher, F. 1996. *Sydowia*, 48, 152.

Reddy, P. V., Bergen, M. S., Patel, R. and White, J. F. 1998. *Mycologia*, 90, 108. pp. 31–65.

Reis, V. M., Baldani, J. I., Baldani, V. L. D. and Dobereiner, J. 2000. *Crit. Rev. Plant Sci.*, 19, 227.

Reiter, B., Pfeifer, U., Schwas, H. and Sessitsch, A. 2002. *Applied and Environmental Microbiology*, 68 (5), 2261-2268.

Rodrigue, K. F. and Samuls, G. J. 1999. *J. Basic Microbiol.*, 39, 131.

Schena, L., Nigro, F., Pentimone, I., Ligorio, A. and Ippolito, A. 2003. *Postharvest Biol. Technol.*, 30, 209.

Schulz, B. U., Draeger, S. and Aust, H. J. 1993. *Mycol. Res.*, 97, 1447.

Scott, B. 2001. *Curr. Opin. Mcriobiol.*, 4, 393.

Sessitsch, Reiter, B., Pfeifer, U. and Wilhelm, E. 2002. *FEMS Microbiol. Ecol.*, 39, 23.

Siegel, M. R. and Latch, G. C. M. 1987. *Annu. Rev. Phytopathol.*, 25, 293.

Silva, G. A., Lumini, E., Maia, L. C., Bonfante, P. and Bianciotto, V. 2006. *Mycorrhiza*, 16, 183-189.

Smith, H., Wingfield, M. J. and Petrini, O. 1996. *For. Ecol. Manage.*, 89, 189.

Stanley, S. J. 1992. *Can. J. Bot.*, 70, 2089.

Strobel, G. A., Stierle, A., Stierle, D. and Hess, W. M. 1993. *Mycotaxon*, 47, 71.

Strobel, G. and Daisy, B. 2003. *Microbiol. Mol. Biol. Rev.*, vol. 67, pp. 491–502.

Strobel, G., Daisy, B. and Castillo, U. 2005. *Plant Pathology Journal*, 4 (2), 161-176.

Strobel, G., Daisy, B., Castillo, U. and Harper, J. 2004. *J Nat Prod*, 67, 257-268.

Strobel, G., Yang, X., Sears, J., Kramer, R., Sidhu, R. S., Hess, W. M. 1996. *Microbiology*, 142, 435-440.

Strobel, G.A., Microbes and Infection, 2003. 5, 535-544.

Strobel, G.A., Miller, R.V., Miller, C., Condron, M., Teplow, D.B., Hess, W.M. 1999. Cryptocandin, a potent antimycotic from the endophytic fungus *Cryptosporiopsis cf. quercina*, Microbiol 145, 1919–1926.

Sturz, A. V. and Christie, B. R. 1996. *Soil Biol. Biochem.*, 28, 583.

Sturz, A. V., Christie, B. R. and Matheson, B. G. 1998. *Can. J. Microbiol.*, 44, 162.

Sturz, A. V., Christie, B. R., Matheson, B. G., Arsenault, W. J. and Buchanan, N. A. 1999. *Plant Pathol.*, 48, 360.

Sutherland, B. L., Hume, D. E. and Tapper, B. A. 1999. *N. Z. J. Agric. Res.*, 42, 19.

Tan, R. X. and Zou, W.X. 2001. *Nat Prod Repor*, 18, 448-459.

Tejera, N., Ortega, E., Rodes, R. and Lluch, C. 2006. *J. Plant Physiol.*, 163, 80–85.

Wagenaar, M., Corwin, J., Strobel, G. A., Clardy, J. 2000. *J. Nat. Prod*, *63*, 1692-1695.

Walsh, T.A. 1992. Inhibitors of b-glucan synthesis, in: J.A. Sutcliffe, N.H. Georgopapadakou (Eds.), Emerging Targets in Antibacterial and Antifungal Chemotherapy, Chapman & Hall, London, pp. 349–373.

Wilkinson, H. H., Siegel, M. R., Blankenship, J. D., Mallory, A. C., Bush, L. P. and Schardl, C. L. 2000. *Mol. Plant-Microbe Interact.*, 13, 1027.

Young, D. H., Michelotti, E. J., Sivendell, C. S., Krauss. N. E. 1992. *Experientia*, 48, 882-885.

Zhang, W., Wendel, J. F. and Clark, L. G. 1997. *Molecular Phylogenetics and Evolution*, 8, 205–217.

Zou, W. X. and Tan, R. X. 1999. *Adv. Plant Sciences*, 2, 183.

Zou, W. X., Meng, J. C, Lu, H., Chen, G. X, Shi, G. X., Zhang, T. Y., Tan, R. X. 2000. *J. Nat. Prod*, *63*, 1529-1530.

Zuccaro, A., Schulz, B. and Mitchell, J. 2003. *Mycological Research*, 107, 1451–1466.

□□□

Index

R

S

T

❏❏❏